Quantum Reality

Quantum Reality

Shaping The Unseen World

Alina Hazel

UNIEK ENTERPRISES

CONTENTS

INDEX

Introduction

Chapter 1: Quantum Reality
1.1 Setting the stage for the exploration of quantum reality
1.2 Historical overview of quantum mechanics
1.3 Foundational principles and key concepts

Chapter 2: Superposition and Wave-Particle Duality
2.1 Understanding the concept of superposition
2.2 Delving into wave-particle duality
2.3 Experiments and observations revealing these quantum phenomena

Chapter 3: Entanglement and Spooky Action
3.1 Unraveling the mysteries of entanglement
3.2 Experiments that showcase entanglement's strange behavior
3.3 The implications of entanglement on our understanding of reality

Chapter 4: The Uncertainty Principle and Indeterminacy
4.1 Heisenberg's Uncertainty Principle and its significance
4.2 The limits of precision in quantum measurements
4.3 The philosophical implications of indeterminacy

Chapter 5: Quantum Mechanics and the Nature of Space and Time
5.1 Quantum mechanics and its influence on our perception of space and time
5.2 Relativity and quantum physics: Bridging the gap
5.3 The search for a unified theory of physics

Chapter 6: Quantum Mechanics and the Universe
6.1 Quantum cosmology and the early moments of the universe
6.2 The role of quantum reality in the evolution of the cosmos

INDEX

In the terrific venue of the universe, there exists a phase whereupon the most confounding and bewildering execution unfurls — an unpredictable dance between particles, waves, and probabilities. This stage is, as a matter of fact, the universe of quantum mechanics, a domain that challenges our most basic instincts and powers us to reconsider the actual idea of reality itself. Quantum mechanics is a hypothetical system that has, throughout the last hundred years, changed how we might interpret the universe, disentangled the secrets of the tiny, and brought forth innovative wonders that have reshaped our lives.

This excursion into the core of quantum reality starts with the straightforward inquiry of what lies under the surface for the universe. In the old style perspective of physical science, tracing all the way back to the hour of Isaac Newton, the response was clear: matter was made out of particles, and the collaborations between these particles were administered by deterministic regulations. This perspective, frequently alluded to as traditional mechanics, filled in as a dependable starting point for grasping the way of behaving of naturally visible items, from the delicate fall of an apple to the glorious circles of planets in our planetary group.

In any case, as the twentieth century unfolded, breaks started to show up in the façade of traditional material science. Tests at the nuclear and subatomic scales, including electrons, photons, and other basic particles, uncovered ways of behaving that resisted traditional clarification. These deviations from the normal standards were the main clues that something phenomenal was brewing, prowling in the shadows of the quantum domain.

In 1900, the German physicist Max Planck moved into this perplexing world when he presented the idea of quantization. Planck recommended that energy, recently remembered to be nonstop and unendingly separable, could exist in discrete, quantized units. This progressive thought brought forth quantum hypothesis, a system that laid the preparation for the ensuing quantum unrest.

The year 1905 denoted one more significant second in the excursion toward figuring out the quantum world. Albert Einstein, a generally obscure patent inspector at that point, distributed his noteworthy work on the photoelectric impact. In this paper,

Einstein proposed that light is certainly not a consistent wave however is rather made out of discrete parcels of energy, which he called photons. This strong declaration made sense of the photoelectric impact as well as laid out the crucial idea of quantized energy, in concurrence with Planck's prior work.

Einstein's photoelectric impact paper denoted the introduction of quantum mechanics, however the hypothesis really made its mark during the 1920s with the endeavors of a few splendid physicists, most strikingly Niels Bohr, Werner Heisenberg, Erwin Schrödinger, and Max Conceived.

These trailblazers, working freely and cooperatively, fostered a thorough structure for grasping the way of behaving of subatomic particles. This system, presently known as quantum mechanics, stays one of the best and exact logical hypotheses at any point concocted.

Fundamental to the advancement of quantum mechanics is the guideline of wave-molecule duality. As per this standard, particles, for example, electrons show both wave-like and molecule like properties. This duality challenges our traditional instincts, as particles can act as waves, with their positions and momenta becoming unsure and probabilistic. The significant ramifications is that we can never again exactly foresee the position and force of a molecule at the same time, an impediment typified by Heisenberg's renowned vulnerability rule.

Quantum mechanics presents the thought of superposition, where particles can exist in different states at the same time. Schrödinger's condition, a key condition in quantum mechanics, depicts how the quantum condition of a framework develops over the long haul. This condition catches the substance of superposition, permitting particles to be in a straight blend of various states, as opposed to being restricted to a solitary, distinct state. The superposition rule is the way to understanding peculiarities like impedance and the way of behaving of particles in quantum frameworks.

One of the most baffling parts of quantum mechanics is snare. At the point when two particles become entrapped, their properties become connected such that opposes old style clarifications. In any event, when isolated by tremendous distances, the estimation of one trapped molecule momentarily impacts the condition of the other. Einstein broadly alluded to this peculiarity as "creepy activity a ways off." Ensnarement challenges our old style thoughts of causality and proposes a significant interconnectedness in the quantum world.

Quantum mechanics additionally presents the idea of likelihood amplitudes, addressed by complex numbers. These amplitudes, when squared, give the likelihood of a molecule being in a specific state. This probabilistic nature of quantum mechanics is a takeoff from traditional determinism and mirrors the inborn vulnerability that plagues the quantum domain.

While quantum mechanics is phenomenally effective at depicting the way of behaving of particles at the subatomic scale, it stays a hypothetical structure with significant philosophical ramifications. The topic of how this hypothesis affects the idea of reality itself has started many years of discussion among physicists, thinkers, and researchers.

At the core of this discussion is the translation of quantum mechanics. A few translations have been proposed throughout the long term, each offering an alternate point of view on the fundamental idea of the quantum world. The Copenhagen understanding, created by Niels Bohr and Werner Heisenberg, places that the quantum world is innately probabilistic and that the demonstration of estimation implodes the quantum state into a clear result. This translation underlines the job of the eyewitness in characterizing reality, a thought that has prompted a lot of philosophical consideration.

The many-universes understanding, proposed by Hugh Everett III, adopts a fundamentally unique strategy. As per this view, each quantum estimation prompts a spreading of the universe into numerous equal real factors, each comparing to an alternate conceivable result. In the many-universes translation, all prospects coincide in a tremendous multiverse, an idea that challenges our instinct yet gives an answer for the estimation issue in quantum mechanics.

Different translations, for example, the pilot-wave hypothesis and the goal breakdown models, offer elective points of view on quantum reality, each with its own arrangement of assets and shortcomings. The continuous mission to interpret the real essence of quantum reality has led to a rich embroidery of thoughts and ways of thinking, exhibiting the significant and persevering through effect of quantum mechanics on how we might interpret the universe.

The investigation of quantum mechanics stretches out past hypothesis and reasoning, as it has yielded an abundance of commonsense applications. Quantum innovation, specifically, has arisen as a noteworthy boondocks with the possibility to reshape enterprises and our regular routines. Quantum registering, for example, holds the commitment of upsetting computational power by outfitting the one of a kind properties of quantum bits, or qubits. These qubits, in contrast to old style bits, can exist in superpositions and entrapped states, empowering quantum PCs to handle complex issues that are past the range of traditional partners.

Quantum cryptography, one more aspect of quantum innovation, offers remarkable security using quantum key dissemination. By utilizing the standards of entrapment and the no-cloning hypothesis, quantum cryptography gives solid encryption strategies that can possibly shield our advanced correspondence in an undeniably interconnected world.

Besides, quantum sensors and metrology vow to upgrade our capacity to quantify and grasp the actual world with lovely accuracy. Quantum sensors can distinguish minute varieties in actual properties, making them priceless devices in fields like geophysics, clinical diagnostics, and natural observing. The effect of these progressions reaches out to areas as different as medical care, money, and safeguard.

The investigation of quantum mechanics has likewise divulged significant associations between the quantum world and the perceptible universe. Quantum ensnarement has enlivened tests exhibiting the infringement of Chime imbalances, which affirm the presence of non-neighborhood connections between's entrapped particles. These investigations challenge our old style ideas of region and recommend that quantum peculiarities might have outcomes at scales recently remembered to be past the compass of quantum impacts.

In the domain of cosmology, quantum mechanics converges with how we might interpret the universe's initial minutes. The enormous microwave foundation radiation, a remnant of the Huge explosion, gives a window into the universe's earliest stages.

Quantum changes in the early universe left engraves on this radiation, forming the dissemination of systems and vast design we notice today. The interaction between quantum mechanics and cosmology brings up interesting issues about the essential idea of the universe and its starting point.

The significant effect of quantum mechanics isn't restricted to the domain of physical science and innovation. It expands its venture into the domains of reasoning, brain science, and even otherworldliness. The profound associations between the eyewitness and the noticed, as recommended by the Copenhagen translation, have driven some to investigate the ramifications of cognizance in the quantum world. While the possibility of cognizance influencing quantum reality stays a subject of discussion, it features the extensive and interdisciplinary nature of quantum examinations.

Chapter 1

Quantum Reality

The universe of quantum physical science, a domain that challenges our ordinary instinct, has been a wellspring of interest and marvel for researchers and logicians the same. At the core of this cryptic field lies the idea of quantum reality, a system that challenges our regular comprehension of the idea of the universe. In this investigation, we will travel into the profundities of quantum reality, endeavoring to get a handle on the curious and mind-twisting peculiarities that describe this bizarre world.

Quantum mechanics, a hypothesis that arose in the mid twentieth hundred years, gives the establishment to how we might interpret the quantum domain. It was created as a reaction to the restrictions of old style material science in making sense of the way of behaving of particles at the nuclear and subatomic scales. Old style material science had served us well for quite a long time, offering a deterministic and instinctive perspective on the universe. Nonetheless, as researchers tested further into the minuscule domain, they experienced peculiarities that couldn't be made sense of by old style standards.

At the core of quantum mechanics is the wave-molecule duality, a key idea that challenges our old style instincts. As per this standard, particles like electrons and photons show both wave-like and molecule like properties. All in all, they can exist in a condition of superposition, where they are in various places or states all the while, until noticed. This thought, broadly exemplified in Schrödinger's feline Catch 22, is a characterizing component of quantum reality.

One of the essential conditions of quantum mechanics is Schrödinger's wave condition, which portrays the development of quantum frameworks after some time. This condition, albeit profoundly effective in foreseeing the way of behaving of quantum particles, is probabilistic in nature. It gives a probability of tracking down a molecule in a specific state yet can't exactly decide its future state. This probabilistic part of quantum mechanics is an obvious takeoff from the determinism of old style physical science.

The probabilistic idea of quantum mechanics brings up issues about the job of eyewitnesses in the quantum world. The demonstration of estimation or perception implodes the wave capability, deciding the particular condition of a quantum framework. This suggests that reality in the quantum domain is dependent upon the demonstration of perception. The popular twofold cut try shows this idea strikingly. At the point when a molecule, like an electron, is sent through a twofold cut, it acts as both a wave and a molecule until noticed. The simple demonstration of perception falls the wave capability and powers the molecule to take on an unequivocal position.

This carries us to the captivating idea of ensnarement, a peculiarity that Einstein broadly alluded to as "creepy activity a ways off." Trap happens when at least two particles become related so that the condition of one molecule immediately impacts the condition of the other, no matter what the distance isolating them. This non-nearby association challenges our traditional comprehension of causality and proposes that data can travel quicker than the speed of light. While it doesn't consider reasonable correspondence, the ramifications of trap are significant, as it features the interconnectedness of all particles in the quantum domain.

The quantum domain is likewise portrayed by the Heisenberg Vulnerability Standard, which declares that it is difficult to know both the position and force of a molecule with full confidence at the same time. This inborn vulnerability isn't because of constraints in estimation however is a key property of the quantum world. It suggests that in the quantum domain, there are intrinsic cutoff points to what we can be familiar with a molecule, and the demonstration of estimation itself upsets the framework being noticed.

One more key part of quantum the truth is the idea of quantization, which confines specific actual properties to discrete qualities. For instance, electrons can exist in unambiguous energy levels inside an iota, and the energy they transmit or retain is quantized. This quantization of energy levels is answerable for the discrete unearthly lines saw in the light transmitted by particles. It underlines the discrete and granular nature of the quantum world, as an unmistakable difference to the persistent and smooth properties of old style material science.

The basic particles that make up the quantum world are represented by a bunch of quantum numbers that depict their properties. These quantum numbers, for example, the main quantum number, precise force quantum number, and attractive quantum number, decide the energy, shape, and direction of electron circles inside a molecule. Understanding these quantum numbers is vital for anticipating the way of behaving of electrons and the construction of iotas.

One of the most commended utilizations of quantum mechanics is the improvement of quantum PCs. Not at all like traditional PCs, which depend on bits (0s and 1s) to handle data, quantum PCs use qubits, which can exist in different states at the same time because of superposition. This property permits quantum PCs to play out particular sorts of estimations dramatically quicker than traditional PCs. Quantum registering holds the commitment of changing fields like cryptography, enhancement, and material science.

In the journey to comprehend quantum reality, physicists have created different translations of quantum mechanics, each offering an alternate point of view on the idea of the quantum world. One of the most notable translations is the Copenhagen understanding, which underscores the job of the spectator in the breakdown of the wave capability. Another translation is the Many-Universes understanding, which sets that each conceivable result of a quantum occasion happens in a different part of the universe. These translations are the subject of continuous discussion and investigation inside established researchers.

The idea of quantum reality reaches out past the domain of particles and wave capabilities. It likewise has significant ramifications for how we might interpret the perceptible world. Quantum mechanics has been effectively applied to make sense of the way of behaving of intricate frameworks, for example, superconductors and superfluids. It has likewise prompted the improvement of quantum field hypothesis, which depicts the key powers of the universe, including electromagnetism, the frail atomic power, and the solid atomic power.

Quantum field hypothesis brings together the standards of quantum mechanics with the hypothesis of extraordinary relativity, bringing about a structure that records for the way of behaving of particles at relativistic velocities. It has been profoundly fruitful in foreseeing and making sense of the way of behaving of particles in high-energy molecule gas pedals, like the Huge Hadron Collider (LHC). The revelation of the Higgs boson, a molecule related with the Higgs field, was a critical victory of quantum field hypothesis and affirmed the presence of this principal field that grants mass to particles.

The quantum world isn't bound to the research facility or the domain of hypothetical material science. Quantum advancements are progressively tracking down their direction into our regular routines. Quantum cryptography, for instance, utilizes the standards of quantum mechanics to make secure correspondence channels that are basically impenetrable to snooping. Quantum sensors and imaging methods offer remarkable degrees of accuracy and awareness, with applications going from clinical diagnostics to natural observing.

In the field of quantum correspondence, quantum trap assumes a focal part in the improvement of quantum key dispersion (QKD) frameworks. These frameworks empower secure correspondence by taking advantage of the properties of snared particles. Any endeavor to catch the quantum key would upset the ensnarement, making the clients aware of potential listening in. This degree of safety is difficult to accomplish with traditional encryption techniques, making QKD a basic instrument for protecting delicate data.

Quantum processing, as referenced prior, is one more area of fast headway. Organizations and examination establishments are effectively chasing after the improvement of pragmatic quantum PCs that can take care of intricate issues past the range of old style PCs. These incorporate advancement issues, cryptography, drug revelation, and materials science. Quantum calculations can possibly alter businesses and change our way to deal with critical thinking.

The idea of quantum instant transportation, frequently connected with sci-fi, is a genuine peculiarity in the realm of quantum mechanics. Quantum instant transportation includes the exchange of the quantum condition of one molecule to another, frequently over a critical distance. While it doesn't involve the prompt transportation of issue, it is a momentous accomplishment with regards to quantum data move. Quantum instant transportation has functional applications in quantum correspondence and quantum processing.

The standards of quantum reality additionally stretch out to the domain of quantum thermodynamics. Conventional thermodynamics, represented by old style physical science, depicts the way of behaving of naturally visible frameworks. Quantum thermodynamics, then again, investigates the thermodynamic properties of quantum frameworks. It uncovers interesting peculiarities, for example, quantum heat motors, which can work at high productivity by tackling quantum impacts.

Quantum ensnarement and superposition, crucial highlights of quantum reality, have prompted the improvement of quantum sensors with uncommon accuracy. Quantum sensors can quantify a great many actual properties, including attractive fields, gravitational waves, and, surprisingly, the World's pivot. These sensors have pragmatic applications in geophysics, route, and basic physical science research.

1.1 Setting the stage for the exploration of quantum reality

The excursion into the entrancing space of quantum reality starts with a verifiable scenery saturated with logical request and scholarly interest. To appreciate the intricacies and ramifications of the quantum world, one should grasp the crucial change in logical idea that unfurled during the mid twentieth hundred years.

The underpinning of traditional physical science, laid out by illuminators like Newton, Maxwell, and Einstein, gave a strong system to figuring out the actual universe. Newton's laws of movement, Maxwell's conditions of electromagnetism, and Einstein's hypothesis of relativity ruled the logical scene and offered a deterministic point of view on the way of behaving of issue and energy.

Nonetheless, as researchers wandered further into the tiny domains of iotas and subatomic particles, they experienced peculiarities that challenged the standards of traditional physical science. Old style mechanics attempted to make sense of the way of behaving of these little elements, inciting the requirement for another structure to clarify their particular nature.

The introduction of quantum mechanics proclaimed a change in perspective in how we might interpret the actual world. The underpinnings of this momentous hypothesis arose out of the spearheading works of researchers like Max Planck, Albert Einstein, Niels Bohr, Werner Heisenberg, Erwin Schrödinger, and others. Planck's presentation of the quantum speculation in 1900, suggesting that energy is quantized in discrete units or 'quanta,' laid the foundation for the quantum transformation.

Einstein's clarification of the photoelectric impact in 1905, which showed the molecule like nature of light (photons), further tested traditional wave hypotheses of light. This laid the basis for the idea of wave-molecule duality, which turned into a foundation of quantum mechanics.

Bohr's model of the particle, enlivened by the quantization of rakish energy, presented the idea of quantized energy levels for electrons circling the nuclear core. This model effectively made sense of the discrete unearthly lines saw in nuclear emanation and retention spectra, denoting a takeoff from the persistent energy changes anticipated by old style material science.

The plan of Heisenberg's Vulnerability Guideline in 1927, a critical commitment to quantum mechanics, acquainted a principal limit with the accuracy with which certain sets of actual properties, like position and energy, could be all the while known. This rule broke the thought of determinism, proposing intrinsic vulnerabilities in the actual texture of the quantum world.

Schrödinger's wave condition, created around a similar time, gave a numerical structure to portray the way of behaving of quantum frameworks. This condition, represented by complex numbers and wave capabilities, offered a probabilistic portrayal of the way of behaving of particles at the quantum level. It depicted the development of quantum frameworks over the long haul and empowered expectations about the probability of tracking down particles in unambiguous states or positions.

One of the most charming parts of quantum mechanics is the standard of superposition, where particles, for example, electrons can exist in various states all the while until noticed. This idea challenges our old style instinct, proposing that particles can display both wave-like and molecule like properties simultaneously. Schrödinger's renowned psychological study including a feline in a condition of superposition embodies the confounding idea of this rule.

The foundation of quantum the truth is the job of perception in falling the wave capability. As per quantum mechanics, the demonstration of estimation or perception decides the particular condition of a quantum framework. This infers that the truth of the quantum world is dependent upon the demonstration of perception, bringing up significant issues about the idea of reality itself.

The famous twofold cut explore, a foundation of quantum mechanics, strikingly shows the wave-molecule duality and the job of perception. At the point when particles, for example, electrons are sent through a twofold cut, they display wave-like way of behaving and produce an impedance design on the finder. Notwithstanding, when the particles are noticed or estimated, they act like particles, and the impedance design vanishes, featuring the impact of perception on the way of behaving of quantum substances.

Entrapment, one more confusing part of quantum reality, alludes to the connection between's particles with the end goal that the condition of one molecule promptly impacts the condition of another, no matter what the distance between them.

This non-neighborhood association challenges traditional thoughts of region and recommends that particles can be interconnected in manners that rise above spatial detachment.

The EPR conundrum, proposed by Einstein, Podolsky, and Rosen in 1935, featured the obvious dumbfounding ramifications of trap. It focused on the apparently quick correspondence between entrapped particles, an idea Einstein broadly alluded to as "creepy activity a good ways off."

John Chime's hypothesis during the 1960s gave a strategy to test the legitimacy of quantum snare and its non-neighborhood connections. Ensuing trials, for example, the Angle explore, upheld the expectations of quantum mechanics, inclining toward the presence of non-nearby connections between's trapped particles.

The investigation of quantum reality stretches out past the infinitesimal domain and has significant ramifications for how we might interpret the naturally visible world. Quantum mechanics has been effectively applied to make sense of the way of behaving of mind boggling frameworks, for example, superconductors and superfluids. It has prompted the advancement of quantum field hypothesis, which depicts the key powers of the universe, including electromagnetism, feeble atomic power, serious areas of strength for and force.

Quantum field hypothesis brings together quantum mechanics with the hypothesis of exceptional relativity, giving a system to grasping the way of behaving of particles at relativistic paces. It has been instrumental in anticipating and making sense of the way of behaving of particles in high-energy molecule gas pedals, like the Enormous Hadron Collider (LHC). The disclosure of the Higgs boson, a molecule related with the Higgs field, was a huge victory of quantum field hypothesis and affirmed the presence of this essential field that bestows mass to particles.

The investigation of quantum reality has prompted the improvement of different translations of quantum mechanics, each offering an alternate point of view on the idea of the quantum world. The Copenhagen understanding underlines the job of the spectator in the breakdown of the wave capability, setting that estimation makes reality. Then again, the Many-Universes translation proposes that each conceivable result of a quantum occasion happens in a different part of the universe.

The idea of quantum reality stretches out past the research facility and hypothetical material science. Quantum innovations are continuously becoming coordinated into our regular routines. Quantum cryptography, for example, use the standards of quantum mechanics to make secure correspondence channels that are essentially impenetrable to listening in. Quantum sensors and imaging procedures offer phenomenal degrees of accuracy and responsiveness, with applications going from clinical diagnostics to ecological checking.

The rise of quantum processing remains as one of the most encouraging uses of quantum mechanics. Dissimilar to traditional PCs, which depend on bits (0s and 1s) to handle data, quantum PCs use qubits, which can exist in different states at the same time because of superposition. This property permits quantum PCs to play out particular kinds of computations dramatically quicker than traditional PCs, promising huge headways in fields like cryptography, enhancement, and material science.

The philosophical ramifications of quantum mechanics proceed to interest and challenge how we might interpret reality, time, and cognizance. The idea of estimation and perception, the idea of the "quantum mind," and the non-straight nature of time in the quantum domain are subjects of progressing consideration and discussion.

1.2 Historical overview of quantum mechanics

Quantum mechanics is a progressive logical hypothesis that has significantly formed how we might interpret the central structure blocks of the universe. Its advancement can be followed back to the late nineteenth and mid twentieth hundreds of years, as physicists wrestled with the restrictions of traditional material science and tried to make sense of the way of behaving of issue and energy at the littlest scales. This authentic outline of quantum mechanics will take us through the key achievements, thoughts, and figures that have added to the detailing of this historic hypothesis.

The underlying foundations of quantum mechanics can be found in the late nineteenth century when researchers were investigating the idea of light. At that point, the common hypothesis, known as old style material science, depicted light as a persistent wave. In any case, this hypothesis confronted a few difficulties in making sense of specific peculiarities, like the photoelectric impact, where electrons are radiated from a material when presented to light. Traditional physical science anticipated that the energy of the produced electrons ought to rely upon the power of the light, however trial results showed that it relied upon the recurrence or shade of the light. This inconsistency brought up issues about the precision of traditional physical science and established the groundwork for quantum hypothesis.

One of the earliest supporters of the advancement of quantum mechanics was Max Planck. In 1900, Planck presented the idea of quantization, recommending that energy isn't persistent however comes in discrete bundles, or quanta. He recommended that the energy of light is quantized in products of a crucial steady, which is currently known as Planck's consistent. This historic thought gave a clarification to the noticed errors in the photoelectric impact and denoted the introduction of quantum physical science.

Albert Einstein further high level the comprehension of quantum mechanics with his 1905 paper on the photoelectric impact. In this paper, Einstein presented the idea of the photon, a discrete molecule of light, to make sense of the photoelectric impact. He proposed that photons convey energy in discrete units, and their energy is straightforwardly relative to their recurrence.

This work not just offered solid help for the quantization of energy yet in addition established the groundwork for the wave-molecule duality of light, a central idea in quantum mechanics.

One more critical improvement in the early history of quantum mechanics was the Bohr model of the particle, proposed by Niels Bohr in 1913. The Bohr model was a significant takeoff from the old style model of the iota, which portrayed electrons circling the core in a way like planets circling the sun. Bohr's model presented quantized energy levels for electrons in particles. As per his model, electrons could possess explicit energy levels, and advances between these levels brought about the outflow or retention of discrete parcels of energy, known as photons. Bohr's model effectively made sense of the ghastly lines of hydrogen and laid the basis for grasping the electronic construction of iotas.

In any case, the Bohr model had limits, especially in its powerlessness to make sense of the way of behaving of molecules with more than one electron. Subsequently, the quest for a more thorough hypothesis proceeded. It was in this setting that Werner Heisenberg, Max Conceived, and Pascual Jordan created lattice mechanics in 1925, and Erwin Schrödinger formed wave mechanics around the same time. These two numerical definitions of quantum mechanics were at first seen as unmistakable methodologies yet were subsequently demonstrated to be same. Schrödinger's wave condition, specifically, turned into a basic device for portraying the way of behaving of quantum frameworks and is still generally utilized in current quantum material science.

The unification of grid mechanics and wave mechanics, alongside the improvement of the numerical system of quantum mechanics, denoted a huge defining moment throughout the entire existence of this field. The Schrödinger condition, which depicts the time development of quantum frameworks, turned into the foundation of quantum mechanics. It addressed a significant takeoff from traditional material science, where the way of behaving of particles could not entirely settled, and on second thought presented the idea of likelihood amplitudes to depict the way of behaving of particles at the quantum level.

The rule of superposition, a major idea in quantum mechanics, rose up out of this numerical system. It expresses that a quantum framework can exist in a straight mix of different states all the while. This thought was in conflict with old style material science, where a molecule's state could not entirely set in stone, and it brought a degree of indeterminacy into the way of behaving of quantum frameworks. Werner Heisenberg formalized this thought with his renowned vulnerability standard, which expresses that it is difficult to know the specific position and force of a molecule with inconsistent accuracy all the while. The more precisely one knows one of these properties, the less precisely the other still up in the air.

One of the most striking parts of quantum mechanics is its probabilistic nature. Not at all like traditional physical science, which gives deterministic forecasts, quantum mechanics manages probabilities. This inborn vulnerability prompted a change in the manner physicists moved toward the depiction of actual frameworks.

Rather than making deterministic forecasts about a molecule's way of behaving, quantum mechanics permitted researchers to work out the probabilities of various results. This probabilistic nature was represented by the popular psychological test known as Schrödinger's feline, which featured the unusual and unreasonable parts of quantum hypothesis.

The improvement of quantum mechanics was not without its contentions and discussions. The Bohr-Einstein banters during the 1920s were a critical part throughout the entire existence of quantum material science. Albert Einstein, alongside partners Boris Podolsky and Nathan Rosen, tested the culmination and philosophical ramifications of quantum mechanics. They contended that quantum mechanics appeared to infer a sort of "creepy activity a good ways off" or non-region, where the condition of one molecule could immediately influence the condition of another, in any event, when isolated by huge distances. These discussions brought up significant issues about the key idea of the real world and the job of determinism in the quantum world.

Regardless of the discussions and conflicts, quantum mechanics kept on flourishing as a strong and effective system for making sense of the way of behaving of particles at the nuclear and subatomic levels. The hypothesis' prosperity was additionally exhibited by the improvement of quantum electrodynamics (QED) during the 1930s. QED is a quantum field hypothesis that depicts the cooperations between charged particles and electromagnetic fields. It gave unbelievably exact expectations and is viewed as quite possibly of the best hypothesis throughout the entire existence of physical science.

The coming of quantum field hypothesis denoted a change in center from single-molecule quantum mechanics to the investigation of quantum fields and their collaborations. Richard Feynman, Julian Schwinger, and Tomonaga Shinichiro autonomously created various plans of QED, every one of which was subsequently demonstrated to be same. Their work presented the idea of renormalization, which resolved issues connected with vast qualities in the hypothesis and made it conceivable to ascertain physical observables with shocking precision.

The twentieth century considered the development of quantum mechanics to be a complete system for grasping the way of behaving of particles, iotas, and particles. It additionally laid the foundation for the advancement of quantum field speculations, which depict the way of behaving of crucial particles and the major powers of nature. The standard model of molecule material science, planned during the twentieth 100 years, is a finish of these turns of events and gives a bound together depiction of the electromagnetic, powerless, and solid atomic powers, alongside the rudimentary particles that make up the universe.

The standard model is a surprising accomplishment throughout the entire existence of material science. It effectively depicts the way of behaving of subatomic particles, predicts the presence of particles like the Higgs boson, and has been tentatively affirmed to a serious level of accuracy.

It has permitted researchers to investigate and comprehend the key powers and particles that oversee the way of behaving of the universe, and it has made the way for the investigation of extraordinary peculiarities like dim matter and dim energy.

The improvement of quantum mechanics has additionally had significant ramifications for innovation. Quantum hypothesis has brought about various innovative headways, from the advancement of lasers and semiconductors to the field of quantum figuring. Lasers, for instance, work on the standards of quantum mechanics, tackling the quantized energy levels of electrons to create intelligible and extreme light. The semiconductor business, which supports current hardware, depends on the quantum conduct of electrons in materials. Quantum figuring, then again, holds the commitment of reforming calculation by taking advantage of the standards of superposition and entrapment to perform complex computations at speeds impossible by traditional PCs.

One more interesting part of quantum mechanics is the peculiarity of trap. This peculiarity happens when at least two particles become corresponded so that the properties of one molecule are reliant upon the properties of the others, no matter what the distance that isolates them. Albert Einstein broadly alluded to ensnarement as "creepy activity a ways off" and scrutinized its legitimacy. In any case, various trials have since affirmed the presence of entrapment, and it assumes a focal part in quantum innovations like quantum instant transportation and quantum cryptography.

1.3 Foundational principles and key concepts

Fundamental standards and key ideas in any logical discipline act as the structure blocks whereupon the whole field is built. In the domain of quantum mechanics, these standards and ideas are especially fascinating and now and again unreasonable, testing our old style instincts about the way of behaving of the actual world. In this investigation, we will dig into the center standards and key ideas that support the captivating and progressive field of quantum mechanics.

One of the key standards of quantum mechanics is the wave-molecule duality, which attests that particles, for example, electrons and photons can show both wave-like and molecule like properties. This rule emerged from exploratory perceptions that couldn't be made sense of by old style physical science. For instance, the popular twofold cut try exhibited that when particles like electrons or photons are coordinated through two cuts and hit a screen on the opposite side, they make an obstruction design like the way of behaving of waves. This peculiarity recommends that particles can show wave-like way of behaving, including obstruction and diffraction.

Wave-molecule duality is typified in the de Broglie frequency, an idea presented by Louis de Broglie in 1924. It connects a frequency with a molecule's force, proposing that all matter shows wave-like properties.

The frequency is contrarily corresponding to the force of the molecule, suggesting that particles with lower energy (like electrons) have longer related frequencies and, in this way, more articulated wave-like qualities.

Quantum superposition is another fundamental idea that rises up out of wave-molecule duality. It expresses that quantum frameworks can exist in numerous states at the same time. All in all, a quantum molecule, like an electron, can be in a superposition of various states, with each state having a likelihood plentifulness that decides the probability of estimating the molecule in that state. The idea of superposition presents a degree of vulnerability and indeterminacy into quantum mechanics, in a general sense recognizing it from old style material science.

The Schrödinger condition is a focal numerical device in quantum mechanics that depicts the development of quantum frameworks after some time. This condition is the quantum simple of Newton's laws of movement and is utilized to compute the wavefunction of a quantum framework, which encodes the likelihood amplitudes related with various potential states. The Schrödinger condition is a foundation of quantum mechanics, giving a deterministic structure to working out what's to come conditions of quantum frameworks in view of their underlying circumstances.

Be that as it may, the probabilistic idea of quantum mechanics is likewise represented by the Heisenberg vulnerability guideline, which is a vital idea in the field. Formed by Werner Heisenberg in 1927, this standard expresses that it is difficult to know the specific position and force of a molecule with erratic accuracy all the while. The more unequivocally one knows the place of a molecule, the less exactly one can know its energy as well as the other way around. This inherent vulnerability emerges from the wave-molecule duality and is a principal restriction of our capacity to make exact estimations at the quantum level.

Quantum ensnarement is one more idea that has enraptured the creative mind of researchers and the public the same. It alludes to the peculiarity where at least two particles become corresponded so that the properties of one molecule are reliant upon the properties of the others, in any event, when they are isolated by enormous distances. This entrapment can happen through different connections, like the protection of precise energy in a couple of trapped particles. Albert Einstein broadly alluded to trap as "creepy activity a ways off," and it tested traditional ideas of territory and causality.

Ensnarement has been tentatively affirmed through trial of Ringer's hypothesis, named after physicist John Chime, who proposed it during the 1960s. Ringer's hypothesis lays out imbalances that, whenever disregarded in tests, serious areas of strength for give to the presence of ensnarement. Various trials have shown that these imbalances are for sure disregarded, building up the truth of quantum trap.

The idea of quantum estimation is integral to quantum mechanics and is personally associated with wavefunction breakdown. At the point when a quantum framework is estimated, the wavefunction, which encodes the probabilities of different states, falls into one of those states. The demonstration of estimation itself presents a degree of irregularity, and the result can't be anticipated with conviction. This presents the thought of characteristic irregularity in quantum mechanics, rather than old style physical science, where estimations are supposed to yield deterministic outcomes.

The idea of quantum states and observables is a basic part of quantum mechanics. A quantum state portrays the total data about a quantum framework, and it is addressed by a wavefunction. Observables are actual properties or amounts that can be estimated, like position, force, or energy. In quantum mechanics, observables are related with administrators, and the demonstration of estimating a perceptible compares to tracking down the eigenvalues of the comparing administrator. This connection among states and observables is fundamental to grasping the probabilistic idea of quantum estimations.

One of the most captivating parts of quantum mechanics is the idea of quantum burrowing. This peculiarity permits particles to go through energy hindrances that, as per old style material science, ought to be outlandish. Quantum burrowing has a great many applications, from making sense of the atomic combination processes in stars to the activity of passage diodes in hardware. It likewise assumes a pivotal part in the way of behaving of particles in nuclear and sub-atomic frameworks, adding to synthetic responses and the security of issue.

The idea of quantization, presented by Max Planck, is at the core of quantum mechanics. It states that specific actual properties, like energy, are quantized and can take on discrete qualities. For example, the energy levels of electrons in a molecule are quantized, and advances between these levels relate to the emanation or retention of discrete bundles of energy, known as photons. This idea on a very basic level difficulties the consistent idea of traditional physical science and highlights the discreteness of the quantum world.

The probabilistic idea of quantum mechanics is profoundly attached to the idea of likelihood amplitudes. These amplitudes are complicated numbers that encode the probability of a quantum framework being in a specific state. They are utilized to work out the probabilities of different estimation results, and their stages assume a critical part in impedance impacts. Likelihood amplitudes are controlled utilizing numerical tasks, for example, the expansion of amplitudes for various states and the square modulus to get probabilities.

The idea of quantum states is firmly connected to the guideline of unitarity, which expresses that the complete likelihood of all potential results in a quantum estimation should constantly rise to one. This rule mirrors the preservation of likelihood and guarantees that the probabilities of all potential results amount to sureness. Unitarity is a key component of quantum mechanics and is an outcome of the standardization state of the wavefunction.

Quantum mechanics additionally presents the idea of quantum states being portrayed by thickness grids, which are utilized to address blended states. A blended state is a factual group of unadulterated states with related probabilities. Thickness frameworks give a method for portraying quantum frameworks that are not in unadulterated states, and they are fundamental in the investigation of open quantum frameworks and quantum data hypothesis.

The idea of precise force assumes a huge part in quantum mechanics. Precise force is a vector amount that portrays the rotational movement of particles. In quantum mechanics, precise energy is quantized, meaning it can take on specific discrete qualities. The quantization of precise energy is liable for the quantized energy levels of electrons in iotas and is a vital component in grasping nuclear and sub-atomic construction.

The idea of quantization reaches out to other actual properties also. For example, the quantization of charge is an essential guideline in molecule material science, where the electric charge of rudimentary particles is generally a numerous of the rudimentary charge. This rule is fundamental in grasping the way of behaving of particles and their cooperations at the subatomic level.

Quantum mechanics likewise presents the idea of wavefunctions ready and energy space, every one of which gives an alternate point of view on a quantum framework. The wavefunction in place space portrays the probabilities of tracking down a molecule at various situations, while the wavefunction in energy space depicts the probabilities of various force values. These two portrayals are connected through numerical changes, for example, the Fourier change, and give correlative data about a quantum framework.

The standard of evenness is one more key idea in quantum mechanics. Balances assume an essential part in grasping the way of behaving of particles and their collaborations. Evenness standards are intently attached to preservation regulations and give bits of knowledge into the choice guidelines that administer the permitted changes between quantum states. For instance, the protection of rakish energy is related with rotational evenness, and the preservation of electric charge is connected to check balance.

Chapter 2

Superposition and Wave-Particle Duality

Superposition and wave-molecule duality are two of the most significant and captivating ideas in the field of quantum mechanics. These standards challenge old style instincts about the idea of issue and energy, offering another point of view on the key way of behaving of particles at the quantum level.

Superposition is a focal idea in quantum mechanics, and it comes from the wave-like nature of quantum particles. It expresses that quantum frameworks can exist in a direct mix of different states at the same time. All in all, a quantum molecule can possess various states, with each state having a related likelihood sufficiency that decides the probability of estimating the molecule in that state. This idea brings a degree of indeterminacy into quantum mechanics, on a very basic level distinctive it from traditional material science.

The thought of superposition is most broadly exhibited in the twofold cut explore. At the point when particles like electrons or photons are coordinated through two cuts and hit a screen on the opposite side, they make an obstruction design, similar as the way of behaving of waves. This peculiarity proposes that particles can show wave-like way of behaving, like obstruction and diffraction. The impedance design is a consequence of the superposition of probabilities for the molecule to go through one or the other cut and is a sign of quantum conduct.

The numerical structure for depicting superposition depends on the rule of linearity. In quantum mechanics, the wavefunction, which addresses the quantum condition of a framework, is a complex-esteemed capability that encodes the likelihood amplitudes of different states. The rule of linearity expresses that on the off chance that a quantum framework can exist in states An and B, it can likewise exist in a superposition of states An and B, addressed as $\alpha A + \beta B$, where α and β are complicated coefficients that portray the likelihood amplitudes. The probabilities of estimating the framework in states An and B are given by the square extents of these coefficients.

One of the striking elements of superposition is that it considers the chance of obstruction. At the point when at least two quantum states are joined, their likelihood amplitudes can meddle helpfully or damagingly, bringing about particular examples of estimation results. This impedance peculiarity can prompt startling and nonsensical outcomes in quantum tests, testing our traditional instincts about how particles act.

Wave-molecule duality is another fundamental idea in quantum mechanics that further delineates the double idea of particles. It attests that particles, like electrons and photons, can display both wave-like and molecule like properties. This rule emerged from trial perceptions that couldn't be made sense of by old style material science. The way of behaving of particles in quantum mechanics is innately probabilistic and wave-like, and this duality is key to grasping the way of behaving of quantum frameworks.

The wave-molecule duality idea was first presented by Louis de Broglie in 1924 when he recommended that all matter shows wave-like properties. He connected a frequency with particles, known as the de Broglie frequency, which is contrarily corresponding to their energy. This idea recommends that particles with lower energy (e.g., electrons) have longer related frequencies and, hence, more articulated wave-like attributes.

The wave-molecule duality can be seen in different peculiarities and examinations. For example, the diffraction of electrons going through a precious stone cross section shows their wave-like nature. Electrons show obstruction designs like those of light while going through a precious stone cross section, which must be made sense of by thinking about them as waves. This conduct goes against the traditional idea of particles following distinct directions.

Besides, the Compton impact, found by Arthur Compton, gives exploratory proof of the molecule idea of light. In this examination, X-beams occurrence on a material disperse flexibly, and the adjustment of their force and frequency is predictable with the possibility that they act as particles (photons) while communicating with issue.

The wave-molecule duality isn't restricted to electrons and photons. It reaches out to all particles in the quantum domain, testing our traditional instincts about the discrete and distinct way of behaving of issue.

Quantum mechanics compels us to acknowledge that particles can show both wave-like and molecule like properties all the while, contingent upon the exploratory setting.

The idea of wave-molecule duality is intently attached to the probabilistic idea of quantum mechanics. The wavefunction, which portrays the quantum condition of a framework, encodes the likelihood amplitudes of different states. The square of the greatness of the wavefunction at a particular point gives the likelihood of tracking down a molecule at that area. The wavefunction is a numerical portrayal of the molecule's likelihood dispersion and, in this sense, is suggestive of the traditional wave capability used to depict the way of behaving of waves.

With regards to wave-molecule duality, the wavefunction is a complex-esteemed capability that sways in space, like a wave. These motions can prompt the obstruction impacts saw in the twofold cut explore and other quantum tests. The wavefunction epitomizes the wave-like way of behaving of particles, yet when estimated, the molecule shows up as a point-like substance, predictable with the molecule part of its temperament.

The connection between wave-molecule duality and superposition becomes clear while thinking about the way of behaving of quantum particles in a twofold cut explore. As a quantum molecule goes through two cuts, it can exist in a superposition of states — having gone through the left cut and having gone through the right cut. The related likelihood amplitudes of these two states slow down one another, making the impedance design on the screen. This shows that superposition and wave-molecule duality are unpredictably connected, as the superposition of quantum states is fundamental for understanding the wave-like way of behaving of particles.

The idea of wave-molecule duality additionally challenges our old style thoughts of determinism. In traditional physical science, particles follow distinct directions and have exact properties, making their way of behaving actually unsurprising. Interestingly, quantum mechanics presents a degree of inborn vulnerability because of the wave-like nature of particles. The position and energy of a quantum molecule are depicted by likelihood circulations, and their exact qualities can't be all the while known with inconsistent accuracy, as portrayed by the Heisenberg vulnerability rule.

One of the striking results of wave-molecule duality and superposition is the peculiarity of quantum obstruction. Obstruction is a key part of quantum mechanics and emerges when likelihood amplitudes related with various quantum states consolidate. This can prompt intensification or wiping out of specific results, contingent upon the general periods of the likelihood amplitudes.

The obstruction impacts saw in tests feature the probabilistic idea of quantum mechanics. They show that particles don't follow deterministic directions yet rather display a scope of potential results, each with a related likelihood. The way of behaving of quantum frameworks is intrinsically questionable, and the probabilities of various results are determined utilizing the wavefunction and its superposition of states.

Impedance has reasonable ramifications in different areas of quantum innovation. For instance, in quantum figuring and quantum calculations, impedance is bridled to perform estimations and take care of issues more proficiently than old style PCs. Quantum impedance likewise assumes a basic part in quantum cryptography, where it guarantees the security of quantum correspondence conventions.

Besides, the peculiarity of obstruction isn't restricted to twofold cut explores different avenues regarding electrons or photons. It very well may be seen in an assortment of quantum frameworks, including the obstruction of issue waves (e.g., electrons and molecules) and the impedance of quantum fields (e.g., in the activity of lasers). These assorted models outline the pervasiveness of impedance in quantum mechanics and its central job in profoundly shaping the way of behaving of particles and fields.

Wave-molecule duality and superposition additionally have suggestions for the comprehension of wavefunctions and quantum states. The wavefunction depicts the total quantum condition of a framework, and working out the probabilities of various estimation outcomes can be utilized. At the point when a quantum framework is in a superposition of states, its wavefunction is a blend of the wavefunctions of those singular states.

Numerically, the superposition of wavefunctions includes adding the wavefunctions of the singular states with fitting likelihood amplitudes. The square size of the subsequent wavefunction addresses the likelihood appropriation of the quantum framework in its superposition. The idea of wave-molecule duality is implanted in this portrayal, as it mirrors the wave-like person of the wavefunction.

The probabilistic translation of the wavefunction is firmly connected with the idea of estimation in quantum mechanics. At the point when an estimation is made on a quantum framework, the wavefunction falls to one of the potential states, as per the probabilities depicted by the wavefunction. The demonstration of estimation brings a component of irregularity into quantum mechanics, and the result can't be anticipated with assurance, rather than the determinism of traditional physical science.

2.1 Understanding the concept of superposition

Understanding the idea of superposition is essential to getting a handle on the exceptional and intriguing nature of quantum mechanics. Superposition is one of the focal standards of quantum physical science, and it underlies large numbers of the odd and unreasonable peculiarities that happen at the quantum level. To appreciate superposition, one should dive into the core of quantum mechanics and investigate how it shapes the way of behaving of particles and frameworks in the quantum world.

At its center, superposition declares that quantum frameworks can exist in a straight blend of various states at the same time. At the end of the day, a quantum molecule or framework can possess more than one state simultaneously, with each state having a related likelihood sufficiency that decides the probability of estimating the molecule or framework in that state. This idea brings a degree of indeterminacy into quantum mechanics, on a very basic level distinctive it from traditional material science.

The idea of superposition arose out of the wave-like nature of quantum particles. One of the key investigations that shows superposition is the well known twofold cut try. In this trial, particles, for example, electrons or photons are coordinated through two cuts and hit a screen on the opposite side. Shockingly, when these particles go through the cuts and hit the screen, they make an impedance design like the way of behaving of waves. This impedance design proposes that particles can display wave-like way of behaving, like obstruction and diffraction.

The numerical structure for portraying superposition is established in the standard of linearity. In quantum mechanics, the wavefunction is a complex-esteemed capability that addresses the quantum condition of a framework. It encodes the likelihood amplitudes related with various states. The rule of linearity expresses that in the event that a quantum framework can exist in states An and B, it can likewise exist in a superposition of states An and B, addressed as $\alpha A + \beta B$, where α and β are complicated coefficients that portray the likelihood amplitudes. The probabilities of estimating the framework in states An and B are given by the square extents of these coefficients.

The square of the size of the superposed wavefunction depicts the likelihood conveyance of the quantum framework in its superposition. This implies that the likelihood of estimating the framework in a specific not entirely set in stone by the square of the relating coefficient in the superposition. Superposition permits quantum frameworks to exist in an express that is a mix of different potential states, with each state adding to the general likelihood conveyance.

One of the most interesting parts of superposition is that it takes into account the chance of obstruction. At the point when at least two quantum states are joined in a superposition, their likelihood amplitudes can meddle helpfully or horrendously, bringing about unmistakable examples of estimation results. This obstruction peculiarity can prompt unforeseen and outlandish outcomes in quantum tests, testing our traditional instincts about how particles act.

The idea of superposition isn't restricted to basic frameworks. It applies to complex quantum frameworks too, like particles and atoms. On account of particles, the way of behaving of electrons is depicted by superpositions of energy levels. An electron in a molecule can exist in a superposition of various energy states, and this superposition leads to the trademark otherworldly lines saw in nuclear spectra.

Superposition likewise has huge ramifications for quantum figuring, a field that outfits the standards of quantum mechanics to play out specific computations more effectively than old style PCs. In a quantum PC, quantum bits or qubits can exist in superpositions of 0 and 1, empowering the equal handling of data. This ability permits quantum PCs to possibly take care of issues that are immovable for traditional PCs, like calculating enormous numbers and streamlining complex frameworks.

Quantum impedance, which emerges from superposition, assumes a significant part in quantum processing calculations. Quantum calculations exploit impedance impacts to enhance the likelihood of acquiring the right response and stifle mistaken results. The superposition of quantum states in a quantum PC is a secret weapon that separates it from traditional PCs and empowers it to perform computations at speeds that are possibly dramatically quicker for specific issues.

The idea of superposition has suggestions past quantum registering. It is fundamental to quantum data hypothesis, which concentrates on the handling and transmission of data utilizing quantum frameworks. Quantum data hypothesis envelops quantum cryptography, quantum instant transportation, and quantum key dissemination, all of which depend on the standards of superposition and ensnarement to get and communicate data in clever ways.

Superposition likewise assumes a part in quantum correspondence, like quantum instant transportation, where the quantum condition of one molecule is moved to one more far off molecule without a direct actual association. In quantum key conveyance, superposition permits the protected trade of encryption keys in view of the standards of quantum mechanics, guaranteeing that listening in on the correspondence is discernible.

The idea of superposition is personally associated with the wave-molecule duality of quantum particles. This duality states that particles, like electrons and photons, can display both wave-like and molecule like properties. The wave-molecule duality is firmly connected with superposition, as it is the wave-like part of particles that permits them to exist in superpositions of states.

For instance, in the twofold cut explore, particles like electrons or photons show obstruction designs while going through two cuts. The obstruction design is a result of the superposition of probabilities for the molecule to go through one or the other cut. This impedance exhibits the wave-like nature of the particles, which prompts the formation of obstruction borders on the screen.

The wave-molecule duality idea was presented by Louis de Broglie in 1924 when he suggested that all matter shows wave-like properties. He connected a frequency with particles, known as the de Broglie frequency, which is contrarily corresponding to their energy.

This idea proposes that particles with lower energy (e.g., electrons) have longer related frequencies and, accordingly, more articulated wave-like attributes.

The wave-molecule duality is apparent in different peculiarities and tests. The diffraction of electrons going through a gem grid, for instance, exhibits their wave-like nature. Electrons display obstruction designs like those of light while going through a precious stone grid, which must be made sense of by thinking about them as waves. This conduct goes against the old style thought of particles following clear cut directions.

Besides, the Compton impact, found by Arthur Compton, gives exploratory proof of the molecule idea of light. In this examination, X-beams episode on a material dissipate flexibly, and the adjustment of their energy and frequency is steady with the possibility that they act as particles (photons) while collaborating with issue.

Wave-molecule duality challenges our old style instincts about the discrete and obvious way of behaving of issue. It drives us to acknowledge that particles can display both wave-like and molecule like properties all the while, contingent upon the exploratory setting.

With regards to superposition, wave-molecule duality recommends that particles can exist in superpositions of states, where each state relates to various potential directions or ways. The likelihood amplitudes related with these states can impede one another, prompting the trademark obstruction designs saw in quantum tests.

The superposition of quantum states has significant ramifications for the idea of the real world and our view of the quantum world. It challenges traditional thoughts of determinism and definiteness, where particles are supposed to have clear cut properties and directions. In the quantum domain, particles exist in superpositions, and their way of behaving is depicted by likelihood dispersions.

The probabilistic understanding of superposition and wave-molecule duality is a central element of quantum mechanics. At the point when an estimation is made on a quantum framework, the wavefunction, which addresses the quantum state, implodes to one of the potential states in the superposition. The demonstration of estimation brings a component of haphazardness into quantum mechanics, and the result can't be anticipated with assurance, rather than the determinism of traditional material science.

The Heisenberg vulnerability standard is firmly connected with the probabilistic idea of quantum mechanics and is an outcome of the wave-molecule duality and superposition. This standard, figured out by Werner Heisenberg in 1927, states that it is difficult to know the specific position and energy of a molecule with erratic accuracy all the while. The more definitively one knows the place of a molecule, the less exactly one can know its force, as well as the other way around.

The vulnerability standard mirrors the intrinsic vulnerability presented by the wave-like nature of particles in superposition. It forces a key constraint on our capacity to make exact estimations at the quantum level. This vulnerability isn't because of impediments in estimation innovation however is a principal component of the quantum world.

Superposition and wave-molecule duality likewise have suggestions for the idea of quantum states and observables. A quantum state depicts the total data about a quantum framework.

2.2 Delving into wave-particle duality

Digging into wave-molecule duality is an excursion into the core of quantum mechanics, a domain where particles, for example, electrons and photons challenge traditional instincts by displaying both wave-like and molecule like properties. Wave-molecule duality is quite possibly of the most significant and fascinating idea in quantum material science, and it challenges how we might interpret the major idea of issue and energy.

At its center, wave-molecule duality affirms that particles, especially those at the quantum level, can display both wave-like and molecule like qualities. This idea is established in the key standards of quantum mechanics and has extensive ramifications for the way of behaving of particles and frameworks in the quantum world.

The historical backdrop of wave-molecule duality can be followed back to the mid twentieth century when researchers wrestled with baffling trial results that couldn't be made sense of by traditional material science. One of the critical figures in the advancement of this idea was Louis de Broglie, who proposed in 1924 that matter, similar to light, has both molecule and wave-like properties. De Broglie connected a frequency with particles, known as the de Broglie frequency, which is conversely relative to their force. This frequency proposed that particles with lower energy (like electrons) have longer related frequencies and, subsequently, more articulated wave-like qualities.

De Broglie's speculation tracked down trial support as the diffraction of electrons. In the twofold cut try, electrons going through two cuts make an impedance design on a screen, similar as the obstruction example of light waves going through two cuts. This peculiarity plainly showed that electrons, normally considered particles, can act like waves when exposed to specific trial conditions.

The idea of wave-molecule duality isn't restricted to electrons. It stretches out to an extensive variety of quantum particles, including photons, iotas, and atoms. Understanding this duality requires diving into the major rules that underlie the way of behaving of particles in the quantum domain.

One of the vital parts of wave-molecule duality is the capacity of particles to display obstruction. Obstruction is an essential peculiarity that happens when waves cross-over, and their amplitudes join either helpfully or disastrously. With regards to quantum mechanics, particles display impedance when they exist in superpositions of states, and their likelihood amplitudes slow down one another.

The superposition of quantum states is a focal idea firmly connected to wave-molecule duality. Superposition declares that quantum frameworks can exist in a direct mix of different states at the same time. At the point when a molecule is in a superposition of states, it can slow down itself, making examples of valuable and damaging impedance that influence the probabilities of estimation results.

The twofold cut try fills in as a surprising exhibition of obstruction with regards to wave-molecule duality. At the point when particles, for example, electrons or photons, go through two cuts and hit a screen, they make an obstruction design comprising of exchanging brilliant and dull edges. This obstruction design emerges from the superposition of probabilities for the molecule to go through one or the other cut.

In the twofold cut explore, the molecule's wavefunction, which addresses its quantum state, exists in a superposition of states comparing to going through the left cut and the right cut. The likelihood amplitudes related with these states can obstruct one another. At the point when these likelihood amplitudes helpfully meddle, they upgrade the likelihood of tracking down the molecule at specific areas on the screen, bringing about the splendid edges. Alternately, when they horrendously meddle, they diminish the likelihood of tracking down the molecule at different areas, prompting the dim edges.

The impedance design saw in the twofold cut explore is a striking illustration of wave-molecule duality in real life. It shows that particles, notwithstanding their singular nature, can act as waves, displaying obstruction and diffraction designs. This conduct difficulties old style instincts about particles having obvious directions and ways of behaving.

Wave-molecule duality is firmly connected with the probabilistic idea of quantum mechanics. In quantum physical science, the quantum condition of a molecule is portrayed by a wavefunction, a complex-esteemed capability that encodes the likelihood amplitudes of various states. The square of the size of the wavefunction at a particular point gives the likelihood of tracking down the molecule at that area.

The probabilistic translation of wave-molecule duality stresses the intrinsic vulnerability related with the way of behaving of quantum particles. The wavefunction addresses the likelihood conveyance of the molecule, and that implies that the exact result of an estimation can't be anticipated with conviction. All things considered, quantum mechanics gives the probabilities of various estimation results in view of the wavefunction.

This probabilistic understanding is principal to quantum mechanics and stands as an unmistakable difference to traditional physical science, where the way of behaving of particles is deterministic and unsurprising. In quantum mechanics, particles don't follow obvious directions yet exist in superpositions of states, prompting probabilistic results when estimated.

Wave-molecule duality challenges our traditional thoughts of determinism and definiteness. It drives us to acknowledge that particles can show both wave-like and molecule like properties all the while, contingent upon the trial setting. The idea of superposition, which permits particles to exist in superpositions of states, is firmly entwined with wave-molecule duality, as superpositions lead to the wave-like way of behaving of particles.

One more fundamental idea connected with wave-molecule duality is the Heisenberg vulnerability rule, which was planned by Werner Heisenberg in 1927. This standard is a crucial result of the wave-like nature of particles and their superposition of states. The vulnerability guideline expresses that it is difficult to know the specific position and force of a molecule with erratic accuracy all the while.

The more definitively one knows the place of a molecule, the less unequivocally one can know its force, as well as the other way around. This impediment on concurrent information is an immediate consequence of the wave-molecule duality and the probabilistic idea of quantum mechanics. The position and force of a molecule are depicted by likelihood conveyances, and their exact qualities can't be at the not set in stone with erratic precision.

The Heisenberg vulnerability standard has critical ramifications for quantum estimations. It forces a basic restriction on our capacity to make exact estimations in the quantum world. This impediment isn't because of specialized requirements yet is a major part of the way of behaving of particles with wave-like qualities.

The connection between wave-molecule duality, superposition, and the Heisenberg vulnerability rule becomes obvious while thinking about the way of behaving of particles in quantum frameworks. Particles exist in superpositions of states, displaying wave-like qualities. This superposition prompts a circulation of probabilities for various results, and the wave-molecule duality is clear in the oscillatory idea of the wavefunction.

At the point when an estimation is made on a quantum framework, the wavefunction implodes to one of the potential states in the superposition, as per the probabilities portrayed by the wavefunction. This demonstration of estimation brings a component of irregularity into quantum mechanics, and the result can't be anticipated with assurance, mirroring the probabilistic idea of the quantum world.

Wave-molecule duality and the Heisenberg vulnerability rule challenge our traditional instincts about the discrete and clear cut conduct of issue. They underscore that the way of behaving of particles at the quantum level is intrinsically questionable and wave-like, and the exact upsides of their properties can't be at the same time known with inconsistent accuracy.

The idea of wave-molecule duality additionally has significant ramifications for how we might interpret the idea of quantum states and observables. In quantum mechanics, a quantum state is portrayed by a wavefunction, which encodes the likelihood amplitudes of various states. Observables, then again, are actual properties or amounts that can be estimated, like position, force, or energy.

In the quantum world, observables are related with administrators, and the demonstration of estimating a detectable compares to tracking down the eigenvalues of the comparing administrator. The connection among states and observables is fundamental to figuring out the probabilistic idea of quantum estimations.

Wave-molecule duality challenges the old style thought of particles having distinct properties. All things being equal, quantum states depict the total data about a quantum framework, and the demonstration of estimation uncovers one of the potential results related with the state. This disclosure is intrinsically probabilistic, and the not entirely settled by the probabilities encoded in the wavefunction.

The idea of wave-molecule duality isn't restricted to individual particles. It additionally applies to complex frameworks, like iotas and atoms. With regards to nuclear and sub-atomic material science, the way of behaving of electrons is depicted by superpositions of energy levels, and this superposition brings about the trademark ghostly lines saw in nuclear spectra.

2.3 Experiments and observations revealing these quantum phenomena

Tests and perceptions that uncover quantum peculiarities play had a urgent impact in forming how we might interpret the quantum world. These tests have tested old style instincts and have given undeniable proof to the major standards of quantum mechanics. By investigating a portion of the critical trials and perceptions that have uncovered these quantum peculiarities, we can acquire knowledge into the fascinating and frequently irrational nature of the quantum domain.

Twofold Cut Analysis:

Perhaps of the most notable trial in quantum mechanics is the twofold cut try. It was initially led with light by Thomas Youthful in the mid nineteenth hundred years and later reached out to electrons and different particles. In this examination, a light emission, like electrons or photons, is aimed at a hindrance with two firmly separated cuts. Behind the hindrance, there is a screen that catches the particles that pass through the cuts.

The astonishing outcome is that when particles are sent through the twofold cuts each in turn, they make an obstruction design on the screen, similar as the example delivered by waves. This impedance design is an unmistakable show of wave-molecule duality, demonstrating that particles can display wave-like way of behaving, like obstruction and diffraction.

Wave-molecule duality is clear in the twofold cut analyze on the grounds that particles exist in superpositions of states, having gone through one or the other cut. The likelihood amplitudes related with these states obstruct one another, prompting the trademark impedance design. This trial challenges the traditional idea of particles as discrete, obvious elements and features their wave-like attributes.

Photoelectric Impact:

The photoelectric impact is one more examination that assumed a urgent part in the improvement of quantum mechanics and gave solid proof to the molecule like nature of light. The peculiarity was first seen by Heinrich Hertz and later made sense of by Albert Einstein in 1905.

In the photoelectric impact, light is coordinated onto a metal surface, and electrons are discharged from the metal when presented to the light. The energy of the produced electrons relies upon the recurrence of the occurrence light. Old style wave hypothesis anticipated that rising the power of the light ought to ultimately prompt the outflow of electrons, no matter what the recurrence.

In any case, the exploratory outcomes showed that main light over a specific recurrence (the edge recurrence) could cause electron emanation, no matter what the light's power. This conduct was predictable with the possibility that light comprises of discrete bundles of energy known as photons, each with a particular energy reliant upon its recurrence.

The photoelectric impact gave unquestionable proof to the quantization of energy in quantum mechanics and showed that light has both wave-like and molecule like properties. This examination was fundamental in the improvement of the idea of the photon, an essential quantum substance.

Compton Dissipating:

Compton dissipating, found by Arthur Compton in 1923, is another examination that upheld the molecule like nature of light. In this trial, X-beams are aimed at an objective material, normally electrons. The X-beams disperse off the electrons, and the subsequent dissipated X-beams have a more drawn out frequency than the episode X-beams.

The key perception is that the adjustment of frequency of the dissipated X-beams is relative to the energy of the electrons. This conduct is steady with the possibility that the X-beams act as particles (photons) that crash into the electrons, moving energy and force to them.

Compton dissipating gave exploratory affirmation of the molecule like nature of light, supporting the wave-molecule duality idea. It showed the way that light can interface with issue as discrete bundles of energy, acting as particles.

Harsh Gerlach Investigation:

The Harsh Gerlach explore, directed by Otto Harsh and Walther Gerlach in 1922, is an exemplary trial that uncovers the quantization of precise energy in quantum mechanics. In this examination, a light emission iotas is coordinated through an inhomogeneous attractive field, which applies a power on the particles' attractive minutes.

The astounding outcome is that the light emission is parted into unmistakable spots on a locator screen, as opposed to framing a consistent conveyance true to form in old style material science. Each spot relates to a nuclear state with a quantized rakish energy part along the heading of the attractive field. This quantization of rakish energy is an immediate outcome of the standards of quantum mechanics.

The Harsh Gerlach explore gives trial proof to the quantization of actual properties and shows the way that quantum frameworks can have discrete, quantized values for perceptible amounts.

Youthful's Twofold Cut Examination with Electrons:

The twofold cut try was stretched out to electrons in the mid twentieth hundred years, giving undeniable proof to the wave-molecule duality of particles. In this rendition of the trial, a light emission is aimed at a boundary with two cuts, like the first twofold cut try different things with light.

Electrons, which are generally considered as particles, display impedance designs on the screen behind the cuts, similar as the obstruction design saw with light. This conduct shows that electrons, regardless of their molecule like nature, can likewise display wave-like qualities when in a superposition of states.

Youthful's Twofold Cut Investigation with electrons supports the idea of wave-molecule duality and difficulties traditional instincts about the discrete way of behaving of issue.

Davisson-Germer Examination:

The Davisson-Germer explore, directed by Clinton Davisson and Lester Germer in 1927, is one more significant analysis that gave solid proof to the wave-like nature of electrons. In this trial, a light emission is aimed at a translucent nickel target. The electrons dissipate off the nickel molecules and make a diffraction design on an identifier screen.

The diffraction design saw in the Davisson-Germer explore is like the diffraction design created by X-beams, showing that electrons can display wave-like way of behaving. This investigation gave trial affirmation of the de Broglie speculation, which recommended that matter, including electrons, displays wave-like properties.

Chapter 3

Entanglement and Spooky Action

The domain of quantum material science is a puzzling and mind-twisting field where the laws of traditional physical science separate, and the peculiar and unreasonable standards of quantum mechanics become an integral factor. Among the numerous peculiarities that challenge how we might interpret the universe, quantum trap and the related "creepy activity a ways off" stand apart as two of the most perplexing and entrancing parts of this field. In this paper, we will dive into the profundities of quantum trap, investigating its beginnings, suggestions, and the getting through puzzler of creepy activity.

At the core of quantum snare lies a crucial property of the quantum world known as superposition. Superposition is the possibility that quantum particles, like electrons or photons, can exist in different states at the same time. As opposed to traditional physical science, where an item must be in each state in turn, quantum particles can exist in a mix of states until they are noticed or estimated. This idea, broadly delineated by Schrödinger's feline, has been the wellspring of endless psychological tests and conversations among physicists and logicians.

The peculiarity of quantum trap emerges when at least two particles become corresponded so that their properties are related. This relationship, known as entrapment, implies that the condition of one molecule is associated with the condition of another, regardless of whether they are isolated by huge distances. The particles could be on inverse sides of the universe, yet the demonstration of estimating one molecule momentarily decides the condition of the other, regardless of the distance that isolates them. This peculiarity, first portrayed by Albert Einstein, Boris Podolsky, and Nathan Rosen in their popular EPR paper in 1935, left established researchers in amazement and started serious discussions about the idea of the real world.

Einstein, alongside his teammates, was profoundly awkward with the ramifications of quantum snare. He broadly alluded to it as "creepy activity a ways off," a term that catches the frightful and apparently non-neighborhood nature of trap. Einstein, in his work to maintain the standards of territory and determinism, proposed that there should be covered up factors overseeing the results of quantum estimations, and that the particles were not really trapped. He accepted that the world ought to be reasonable and deterministic, sticking to old style physical science, and the probabilistic, non-deterministic nature of quantum mechanics was, in his view, fragmented.

In any case, in 1964, physicist John Ringer figured out a bunch of numerical disparities, presently known as Chime's imbalances, that could be tried tentatively to decide if ensnarement was a genuine peculiarity or on the other hand on the off chance that it very well may be made sense of by stowed away factors. To the shock of many, resulting tests testing Chime's imbalances reliably affirmed that snare was for sure a veritable, non-neighborhood peculiarity. This outcome shook the underpinnings of old style material science and embraced the peculiar truth of quantum mechanics.

Quite possibly of the most popular examination that tried Chime's imbalances is the Viewpoint try, led by Alain Angle during the 1980s. In this examination, sets of trapped photons were isolated by enormous distances, and estimations of their polarization were made at the same time. The outcomes overcame traditional presumption and gave solid proof to the truth of quantum snare. The creepy activity a ways off, as Einstein had put it, was an affirmed component of the quantum world.

The ramifications of snare and the affirmation of creepy activity have broad ramifications for how we might interpret the universe. One of the most significant parts of ensnarement is its evident infringement of the speed of light, a crucial breaking point in Einstein's hypothesis of relativity. Ensnared particles can impact each other immediately, no matter what the immense distances that different them, and this opposes the rule that no data or impact can travel quicker than the speed of light. Nonetheless, it means quite a bit to take note of that trap doesn't consider quicker than-light correspondence. It just uncovers connections between's the ensnared particles and doesn't give a way to send data.

The idea of trap likewise difficulties our thoughts of traditional causality, where occasions have a reasonable circumstances and logical results. In the quantum domain, estimations of entrapped particles seem to happen without a first reason. This clear absence of a causal connection between the estimations has prompted the improvement of different understandings of quantum mechanics, like the Copenhagen translation and the many-universes translation, each endeavoring to make sense of the idea of quantum reality.

Additionally, ensnarement has functional applications in the field of quantum registering and quantum cryptography. Quantum PCs, which depend on the standards of superposition and trap, can possibly tackle complex issues at speeds that would be unimaginable for traditional PCs.

Quantum cryptography utilizes the standards of trap to make secure correspondence channels that are hypothetically insusceptible to listening in, on account of the quick recognition of any interruption into the entrapped particles.

The idea of quantum snare likewise has captivating philosophical ramifications. It brings up issues about the idea of the real world, the connection between the onlooker and the noticed, and the restrictions of human getting it. The popular physicist Richard Feynman once said, "I want to securely say that no one figures out quantum mechanics." While we have created numerical formalisms to portray and foresee the way of behaving of quantum frameworks, the real essence of the quantum world remaining parts slippery.

One translation of ensnarement that endeavors to reveal insight into its secrets is the idea of non-region. Non-region proposes that ensnared particles are not discrete substances but rather exist as a feature of a bigger, interconnected entirety. In this view, the demonstration of estimation on one molecule promptly impacts the other in light of the fact that they are not genuinely autonomous articles. All things being equal, they are essential for a bound together quantum framework.

This thought isn't without discussion, as it challenges our old style instincts and ideas of separateness. In any case, non-territory has tracked down help in different trial results, and it offers an alternate point of view on the idea of trap and the basic texture of the universe.

Another understanding that has acquired consideration is the idea of stowed away factors, which looks to protect old style determinism while making sense of the evident non-region of entrapment. Secret factors are speculative properties or data that are hidden from us however decide the results of quantum estimations. Notwithstanding, as referenced prior, trial proof, like Ringer's disparities, has generally precluded the presence of stowed away factors as a suitable clarification for quantum trap.

Notwithstanding these translations and trial proof, the secret of ensnarement perseveres. The significant ramifications of snare for how we might interpret the universe and the basic idea of reality keep on enrapturing the personalities of physicists and scholars the same.

Ensnarement likewise brings up issues about the job of cognizance in quantum mechanics. A few understandings, like the Copenhagen translation, propose that the demonstration of estimation implodes the quantum superposition into a clear state, and this demonstration of perception is viewed as central to the interaction. This has prompted philosophical discussions about the idea of awareness and its relationship to the quantum world.

While these conversations are charming, it's fundamental for note that the job of cognizance in quantum mechanics stays a subject of discussion and hypothesis. Most of physicists don't credit an essential job to cognizance in the way of behaving of quantum frameworks, and they view quantum estimations as cooperations between particles that keep the probabilistic laws of quantum mechanics.

Ensnarement and creepy activity additionally have suggestions for how we might interpret the naturally visible world. While quantum impacts are normally connected with minuscule particles on the nuclear and subatomic scale, a few physicists have investigated the chance of stretching out these impacts to bigger, more perplexing frameworks. This field of exploration, known as quantum macroscopicity, means to examine whether trap and other quantum peculiarities can have recognizable results at the plainly visible level.

One illustration of quantum macroscopicity is the investigation of quantum superposition in huge particles or even natural frameworks. A few specialists have proposed trials to test whether huge, complex particles can exist in a superposition of states, like the manner in which electrons can. Assuming such examinations were effective, they could give experiences into the limit between the quantum and old style universes and proposition new viewpoints on the job of quantum peculiarities in organic cycles.

One more fascinating road of exploration includes the chance of caught conditions of naturally visible items. While ensnarement is generally normally connected with particles like electrons or photons, a few researchers have investigated making and noticing entrapment in bigger frameworks, like little mirrors or even little mechanical gadgets. The test is to detach these frameworks from their current circumstance and control their quantum properties to exhibit the presence of trap at bigger scopes.

The investigation of entrapment and creepy activity likewise crosses with the field of quantum data hypothesis. Quantum data hypothesis investigates the manners by which quantum frameworks can be saddled to process and send data. Trap assumes a focal part in this field, as it considers the formation .

3.1 Unraveling the mysteries of entanglement

The quantum world is a domain of significant secret and intricacy, where the central standards of old style material science never again apply. At the core of this confounding space lies perhaps of its most confusing peculiarity: quantum snare. This peculiarity, frequently alluded to as "creepy activity a ways off," has enamored the personalities of physicists and savants for almost a really long period. In this paper, we will investigate the secrets of ensnarement, its authentic turn of events, the difficulties it presents to how we might interpret the universe, and the continuous mission to disentangle its mysteries.

Quantum entrapment is a peculiarity that happens when at least two particles become corresponded so that their properties are related. These particles can be isolated by tremendous distances, yet changes in the condition of one molecule promptly influence the condition of the other, challenging old style ideas of territory and causality. This scary and apparently non-nearby association between ensnared particles has fascinated and puzzled researchers and scholars since its disclosure.

The idea of entrapment is personally connected to the essential property of quantum superposition. In the quantum world, particles, for example, electrons or photons can exist in numerous states all the while, dissimilar to old style objects, which possess a solitary state at some random time. This standard, frequently represented by Schrödinger's well known psychological study including a feline in a superposition of life and demise, challenges our traditional instincts about the real world.

The tale of snare starts with the improvement of quantum mechanics in the mid twentieth 100 years. During the 1920s, another structure for understanding the way of behaving of subatomic particles was laid out by physicists like Niels Bohr, Werner Heisenberg, and Erwin Schrödinger. This system established the groundwork for the weird and nonsensical peculiarities of the quantum world.

One of the vital figures throughout the entire existence of entrapment is Albert Einstein. In 1935, alongside Boris Podolsky and Nathan Rosen, Einstein distributed a popular paper known as the EPR paper, in which they presented the idea of entrapment and brought up significant issues about the culmination of quantum mechanics. The EPR paper tested the probabilistic idea of quantum mechanics and recommended that there should be covered up factors administering the results of quantum estimations, it were not really caught to propose that particles.

Einstein's inspiration was to safeguard the standards of area and determinism, which were fundamental in traditional material science. He accepted that the quantum world ought to be perceived in a manner that stuck to traditional standards, where each occasion had an unmistakable reason and impacts were restricted by the speed of light. For Einstein, the possibility of particles momentarily impacting each other across significant stretches was profoundly disrupting, driving him to depict it as "creepy activity a ways off."

Nonetheless, the discussion started by the EPR paper was not settled until quite a few years after the fact. In 1964, the physicist John Chime figured out a bunch of numerical imbalances, presently known as Ringer's disparities, that could be tried tentatively to decide if entrapment was a genuine peculiarity or on the other hand in the event that it very well may be made sense of by stowed away factors. Ringer's disparities gave an approach to test the odd expectations of quantum trap exactly.

To the astonishment of many, trial of Chime's disparities reliably affirmed the presence of ensnarement and the infringement of these imbalances. The consequences of these tests exhibited that snare was a certified non-nearby peculiarity, testing Einstein's thought of stowed away factors and building up the strange truth of quantum mechanics.

Perhaps of the most renowned examination that tried Ringer's imbalances is the Perspective analysis, led by Alain Viewpoint during the 1980s. In this analysis, sets of entrapped photons were isolated by huge distances, and estimations of their polarization were made all the while.

The outcomes were in distinct inconsistency to the expectations of old style material science and gave unquestionable proof to the presence of quantum snare.

The trial affirmation of trap through trial of Chime's imbalances had broad ramifications for how we might interpret the universe. It tested the essential rule of territory, which directs that impacts can't travel quicker than the speed of light. Snared particles seemed to convey quickly, regardless of the tremendous distances that isolated them.

This clear infringement of the speed of light, a foundation of Einstein's hypothesis of relativity, stays one of the most bewildering parts of trap.

In any case, it is critical to stress that quantum entrapment doesn't give a method for quicker than-light correspondence. While the caught particles display prompt connections, this peculiarity doesn't consider the transmission of data or impact past the speed of light. The correspondence of data is as yet dependent upon the limits forced by the hypothesis of relativity.

The idea of snare likewise difficulties our old style thoughts of causality, where occasions have an unmistakable circumstances and logical results. In the quantum domain, estimations of trapped particles seem to happen without a previous reason. This evident absence of a causal connection between estimations has prompted the improvement of different understandings of quantum mechanics, each endeavoring to make sense of the idea of quantum reality.

In addition, snare has reasonable applications in the field of quantum processing and quantum cryptography. Quantum PCs, which depend on the standards of super-position and ensnarement, can possibly tackle complex issues at speeds that would be unimaginable for old style PCs. Quantum cryptography utilizes the standards of snare to make secure correspondence channels that are hypothetically insusceptible to snoopping, on account of the quick discovery of any interruption into the entrapped particles.

The idea of quantum entrapment likewise has captivating philosophical ramifications. It brings up issues about the idea of the real world, the connection between the spectator and the noticed, and the restrictions of human getting it. The renowned physicist Richard Feynman once said, "I want to securely say that no one figures out quantum mechanics." While we have created numerical formalisms to portray and foresee the way of behaving of quantum frameworks, the real essence of the quantum world remaining parts tricky.

One translation of trap that endeavors to reveal insight into its secrets is the idea of non-area. Non-territory recommends that entrapped particles are not independent substances but rather exist as a feature of a bigger, interconnected entirety. In this view, the demonstration of estimation on one molecule quickly impacts the other in light of the fact that they are not genuinely autonomous items. All things considered, they are important for a bound together quantum framework.

This thought isn't without discussion, as it challenges our traditional instincts and ideas of separateness. Notwithstanding, non-territory has tracked down help in different trial results, and it offers an alternate point of view on the idea of ensnarement and the hidden texture of the universe.

Another translation that has acquired consideration is the idea of stowed away factors, which looks to protect traditional determinism while making sense of the obvious non-territory of trap. Secret factors are speculative properties or data that are hidden from us yet decide the results of quantum estimations. Notwithstanding, as referenced prior, exploratory proof, like Ringer's imbalances, has generally precluded the presence of stowed away factors as a practical clarification for quantum ensnarement.

Notwithstanding these understandings and trial proof, the secret of snare perseveres. The significant ramifications of ensnarement for how we might interpret the universe and the basic idea of reality keep on dazzling the personalities of physicists and logicians the same.

Snare additionally brings up issues about the job of awareness in quantum mechanics. A few understandings, like the Copenhagen translation, recommend that the demonstration of estimation falls the quantum superposition into a clear state, and this demonstration of perception is viewed as essential to the interaction. This has prompted philosophical discussions about the idea of awareness and its relationship to the quantum world.

While these conversations are captivating, it's vital for note that the job of cognizance in quantum mechanics stays a subject of discussion and hypothesis. Most of physicists don't credit a crucial job to cognizance in the way of behaving of quantum frameworks, and they view quantum estimations as connections between particles that keep the probabilistic laws of quantum mechanics.

Entrapment and creepy activity likewise have suggestions for how we might interpret the perceptible world. While quantum impacts are regularly connected with minuscule particles on the nuclear and subatomic scale, a few physicists have investigated the chance of stretching out these impacts to bigger, more mind boggling frameworks. This field of examination, known as quantum macroscopicity, means to research whether entrapment and other quantum peculiarities can have noticeable outcomes at the perceptible level.

One illustration of quantum macroscopicity is the investigation of quantum superposition in huge particles or even natural frameworks. A few specialists have proposed trials to test whether enormous, complex particles can exist in a superposition of states, like the manner in which electrons can. Assuming such trials were effective, they could give bits of knowledge into the limit between the quantum and old style universes and deal new viewpoints on the job of quantum peculiarities in organic cycles.

3.2 Experiments that showcase entanglement's strange behavior

Quantum ensnarement, frequently portrayed as "creepy activity a good ways off," is quite possibly of the most secretive and charming peculiarity in the domain of quantum physical science. The idea of ensnarement, where particles become corresponded so that adjustments of one molecule's state immediately influence the condition of another, challenges old style instincts about the idea of the real world and the restrictions of our comprehension. In this paper, we will investigate a portion of the trials that have been directed to grandstand the peculiar and wondrous way of behaving of entrapment, affirming its presence and uncovering its confounding properties.

The Angle Analysis: Quite possibly of the most renowned trial that gave solid proof to the presence of quantum snare is the Perspective Examination, led by Alain Viewpoint during the 1980s. This examination tried Ringer's disparities, which were planned to recognize the expectations of old style material science and those of quantum mechanics. Viewpoint's earth shattering examination included snared sets of photons, particles of light, that were isolated by enormous distances. The entrapped photons were at the same time estimated for their polarizations, and the consequences of the analysis reliably disregarded the expectations of traditional physical science. This affirmed that trap is a genuine, non-nearby peculiarity, as the estimations on one photon momentarily impacted the other, no matter what the partition between them.

The Deferred Decision Quantum Eraser: The postponed decision quantum eraser explore is a provocative investigation that uncovers the retroactive idea of quantum estimations. In this trial, a trapped photon is sent through a bar splitter, making two snared photons with connected polarizations. One of these photons goes through a twofold cut device, while the other goes to an indicator. At the point when the photon at the finder is estimated, its snared accomplice's way of behaving retroactively changes. Assuming the experimenter decides to eradicate data about what cut the photon went through, the obstruction design returns on the screen, showing that the photon acts as though it went through the two cuts at the same time. This trial shows the way that the demonstration of estimation can influence a molecule's previous way of behaving, an idea that challenges our traditional comprehension of causality.

Quantum Ensnarement with Enormous Particles: While entrapment is frequently connected with little particles like photons, specialists have led analyses to exhibit trap with monstrous particles, like iotas and atoms. These investigations show the way that the standards of entrapment can apply to bigger frameworks. In one analysis, scientists effectively trapped two naturally visible jewels, each containing trillions of carbon molecules. The trap of gigantic particles further difficulties the limit between the quantum and old style universes and features the potential for noticing quantum peculiarities on a naturally visible scale.

Quantum Instant transportation: Quantum instant transportation is an entrancing examination that utilizes entrapment to communicate the quantum condition of one molecule to another molecule, regardless of whether they are isolated by huge distances.

This cycle includes three snared particles: the molecule to be magically transported, an ensnared sets of particles, and a far off spectator. At the point when the far off eyewitness plays out an estimation on one of the trapped particles, it falls the quantum condition of the magically transported molecule onto the other caught molecule, really "transporting" its quantum data. Quantum instant transportation, while not including the actual exchange of issue, exhibits the force of entrapment in communicating quantum data immediately.

Quantum Cryptography: Ensnarement is at the core of quantum cryptography, a field that spotlights on secure correspondence utilizing the standards of quantum mechanics. In quantum key dissemination (QKD), ensnared particles are utilized to make a common mystery key between two gatherings. Any endeavor to block the key by a busybody would disturb the entrapped state and be perceptible by the imparting parties. Quantum cryptography is viewed as hypothetically solid, as it depends on the crucial properties of ensnarement to guarantee the security of correspondence. A few examinations have effectively shown the commonsense use of quantum cryptography in secure information transmission.

Quantum Impedance Tests: Quantum obstruction tests, for example, the twofold cut analyze, feature the wave-molecule duality of quantum particles and the job of trap in deciding their way of behaving. In the twofold cut try, a molecule, like an electron or a photon, is sent through two cuts, making an obstruction design on a screen. Be that as it may, when the way of the molecule is estimated or noticed, the impedance design vanishes, and the molecule acts as a solitary, limited element. The unusual way of behaving of particles in obstruction tests features the job of entrapment in deciding if a molecule acts as a wave or a molecule, contingent upon the presence of which-way data.

Einstein-Podolsky-Rosen (EPR) Examinations: The EPR Catch 22, presented by Albert Einstein, Boris Podolsky, and Nathan Rosen in their 1935 paper, was the establishment for some trials that tried the presence of entrapment and non-area. Tests in view of the EPR mystery, similar to the Angle Trial referenced before, reliably affirmed the non-neighborhood connections between's entrapped particles. These trials gave solid proof against the chance of stowed away factors administering the results of quantum estimations, as proposed by Einstein, and upheld the weird truth of quantum ensnarement.

Quantum Burrowing: Quantum burrowing, albeit not ordinarily depicted as an entrapment try, is an indication of quantum conduct that features the baffling idea of particles. In quantum burrowing, particles can enter energy hindrances that traditional material science would anticipate to be impervious. This peculiarity is a result of the wave-like nature of particles and their trapped way of behaving. Quantum burrowing has useful applications in different fields, including hardware and atomic physical science, and is a demonstration of the flighty way of behaving of quantum particles.

These examinations all in all exhibit the odd and noteworthy way of behaving of ensnarement in the quantum world. They challenge our traditional instincts, grow the limits of how we might interpret the universe, and proposition functional applications in fields like quantum registering and cryptography.

While these analyses have given huge experiences into the idea of entrapment, there are as yet many inquiries and secrets to be disentangled. The translation of trap and its suggestions for how we might interpret the universe keep on being subjects of discussion and continuous exploration. As innovation and our comprehension of the quantum world development, we might uncover much more significant parts of ensnarement and its job in forming the texture of our existence. Snare, with its entrancing and baffling properties, stays a focal subject in the investigation of the quantum universe, welcoming further examination and investigation into the profundities of this puzzling peculiarity.

3.3 The implications of entanglement on our understanding of reality

Ensnarement, a crucial peculiarity of quantum mechanics, has extensive ramifications for how we might interpret reality and the idea of the universe. This secretive and illogical peculiarity, where particles become interconnected so that their properties are associated, challenges old style instincts and prompts significant inquiries regarding the crucial design of the universe. In this exposition, we will investigate the ramifications of snare on how we might interpret reality, the philosophical and logical inquiries it raises, and its effect on the advancement of quantum advancements.

Testing Old style Instincts

The first and most clear ramifications of snare is its test to old style instincts. In old style material science, we are acquainted with the possibility of separateness and freedom. Objects exist in unmistakable states, and their still up in the air by nearby connections. The idea of entrapment opposes this traditional viewpoint, showing the way that particles can become connected in a manner that rises above old style limits.

In a trapped state, changes to one molecule's properties momentarily influence the properties of its ensnared accomplice, regardless of whether they are isolated by tremendous distances. This peculiarity abuses the guideline of region, which directs that impacts can't travel quicker than the speed of light. Ensnarement proposes a degree of interconnectedness that is hard to accommodate with our traditional instincts.

Non-Territory and the Idea of Creepy Activity

The non-area inborn in trap is quite possibly of its most astounding component. The idea of non-region challenges the rule of territory, which has been a fundamental supposition in material science since the times of Isaac Newton. As per the rule of territory, actual cycles at one area ought not be ready to impact processes at another area quicker than the speed of light.

Notwithstanding, snared particles seem to convey promptly, no matter what the partition between them. This evident infringement of the speed of light, a principal fundamental of Albert Einstein's hypothesis of relativity, has prompted the portrayal of trap as "creepy activity a ways off," a term begat by Einstein, Boris Podolsky, and Nathan Rosen in their popular EPR paper.

Non-territory challenges our old style comprehension of causality, recommending that estimations on entrapped particles can happen without an unmistakable circumstances and logical results relationship. This idea has significant ramifications for our origination of the universe's design and the idea of the real world. It brings up issues about the interconnectedness of all things and whether the universe works as a solitary, brought together element instead of an assortment of discrete, free parts.

Interpretational Difficulties

Trap's suggestions have prodded a plenty of translations in the field of quantum mechanics. These translations offer various ways of conceptualizing and make sense of the peculiarity while giving remarkable viewpoints on the idea of the real world.

Copenhagen Understanding: The Copenhagen translation, related with Niels Bohr and Werner Heisenberg, is one of the most broadly shown translations of quantum mechanics. It places that quantum estimations breakdown the wave capability, bringing a molecule from a superposition of states into a distinct state. As per this translation, the demonstration of perception assumes a basic part in deciding the result of a quantum estimation. This thought brings up issues about the job of awareness in quantum mechanics and the idea of reality itself.

Many-Universes Translation: The many-universes understanding, proposed by Hugh Everett, recommends that all conceivable quantum results really happen in equal universes. In this view, the universe constantly parts into different branches with every quantum estimation, prompting a boundless number of equal real factors. While this translation gives a deterministic and predictable perspective, it presents the idea of a boundless multiverse, which challenges our traditional comprehension of the universe's design.

De Broglie-Bohm Hypothesis: The de Broglie-Bohm hypothesis proposes the presence of stowed away factors that guide the way of behaving of quantum frameworks. In this understanding, particles have unequivocal properties, and their obvious irregularity is because of our obliviousness of these secret factors. While this view jelly old style determinism and causality, it has not acquired broad acknowledgment and stays a subject of discussion.

These translations feature the intricacy of the philosophical difficulties presented by entrapment and quantum mechanics overall. Every translation offers an alternate point of view on the idea of the real world, the job of perception, and the connection between the quantum and traditional universes. The continuous discussion encompassing these translations mirrors the significant and mysterious nature of snare and its suggestions for how we might interpret the universe.

Trap and the Idea of The real world

The presence of entrapment welcomes us to reexamine the idea of reality itself. It prompts us to examine whether our traditional instincts concerning separateness and region are on a very basic level defective and whether the universe works in manners that resist old style limits. In this specific situation, snare gives an extraordinary window into the more profound layers of the real world.

The idea of non-area, where ensnared particles show interconnected conduct, proposes that the world may not be made out of independent, confined substances. All things being equal, it indicates the chance of an interconnected and related universe. This interconnectedness challenges the reductionist perspective on the real world, which looks to make sense of intricate peculiarities by separating them into less difficult, detached parts.

Entrapment's non-nearby way of behaving has driven a few physicists and savants to consider the possibility of a "comprehensive" or "trapped" reality, where everything is interconnected and reliant. In such a view, there are no genuinely discrete elements, and the limits among frameworks and items become obscured. This point of view offers a significant reconsidering of the universe, where the entire is more than the amount of its parts.

The philosophical ramifications of trap additionally reach out to the idea of causality. In traditional physical science, causality is viewed as a chain of occasions, where one occasion prompts one more in a deterministic manner. Entrapment challenges this deterministic view, as estimations on trapped particles can happen without an unmistakable circumstances and logical results relationship. This obvious breakdown of causality has prompted conversations about the idea of time and the chance of retroactive impacts.

The deferred decision quantum eraser try represents the retroactive idea of quantum estimations. In this examination, the decision of whether to delete data about which way a molecule took through a twofold cut device influences the molecule's conduct previously, retroactively deciding if it showed wave-like obstruction or molecule like way of behaving. This outcome brings up issues about the idea of causality and the job of perception in forming reality.

Also, the idea of snare difficulties our old style instincts about the job of cognizance in the quantum world. The Copenhagen understanding, for example, proposes that the demonstration of estimation and perception assumes an essential part in falling the quantum wave capability and deciding the result of an estimation. This understanding has prompted philosophical discussions about the connection among cognizance and the quantum world.

While these conversations are captivating, it's vital to take note of that the job of cognizance in quantum mechanics stays a subject of discussion and hypothesis. Most physicists don't credit a key job to cognizance in the way of behaving of quantum frameworks and view quantum estimations as cooperations between particles that observe the probabilistic laws of quantum mechanics.

Pragmatic Applications and Quantum Advances

Snare, with all its philosophical and calculated secrets, likewise has significant ramifications for commonsense applications and the improvement of quantum advancements. Quantum mechanics, which includes ensnarement, has given the establishment to arising advancements that can possibly reform different fields.

Quantum Processing: One of the main mechanical ramifications of entrapment is in the field of quantum registering. Quantum PCs influence the standards of superposition and entrapment to perform complex computations at speeds that would be infeasible for traditional PCs. The special properties of qubits, for example, their capacity to exist in a superposition of states, empower quantum PCs to investigate a tremendous arrangement space all the while, prompting outstanding speedup for specific kinds of issues. Quantum registering has applications in cryptography, materials science, drug revelation, enhancement issues, and numerous different spaces, offering the commitment of another period of computational power.

Quantum Cryptography: Quantum cryptography, which depends on entrapment, offers secure correspondence channels that are hypothetically insusceptible to snooping. In quantum key dispersion (QKD), caught particles are utilized to make a common mystery key between imparting parties. Any endeavor to catch the key would upset the ensnared state and be discernible by the gatherings in question. Quantum cryptography can possibly change information security and shield delicate data from digital dangers, offering a viable answer for the steadily expanding challenge of secure correspondence.

Quantum Instant transportation: Quantum instant transportation is a captivating use of ensnarement. It permits the exchange of the quantum condition of one molecule to another, regardless of whether they are isolated by enormous distances. While quantum instant transportation doesn't include the actual exchange of issue, it shows the force of entrapment in sending quantum data promptly. This innovation has suggestions for secure correspondence and data transmission.

Quantum Detecting and Imaging: Entrapment is likewise being outfit for high-accuracy detecting and imaging applications. Quantum sensors, which depend on entrapped states, can accomplish remarkable degrees of accuracy in different estimations, including those for gravitational waves, attractive fields, and temperature. Quantum imaging methods, for example, quantum-improved imaging and quantum-upgraded lidar, influence snare to outperform the impediments of traditional sensors, offering the potential for forward leaps in fields like medical services, geophysics, and ecological checking.

The advancement of these quantum innovations isn't just reasonable yet in addition extends how we might interpret ensnarement and its part in the quantum world. They show that snare isn't only a philosophical idea however a principal include with certifiable applications.

Challenges and Progressing Exploration

In spite of the commonsense applications and the philosophical experiences that ensnarement gives, many difficulties and secrets remain. The translation of ensnarement and its suggestions for how we might interpret reality keep on being subjects of progressing examination and discussion.

Specialists are endeavoring to drive the limits of quantum peculiarities into the plainly visible world. The investigation of quantum macroscopicity expects to examine whether entrapment and other quantum impacts can have noticeable outcomes at bigger scopes. Tests have investigated the chance of noticing entrapment in naturally visible items, for example, small mirrors or mechanical gadgets, which could offer new experiences into the limit between the quantum and traditional universes.

The investigation of quantum macroscopicity is connected to the crucial inquiry of how the quantum world points of interaction with the old style world. Understanding this connection point is critical for accepting the more profound nature of the real world and overcoming any barrier between the two domains. It provokes us to investigate the likely cross-over and exchange between the quantum and old style depictions of the universe.

Moreover, the improvement of cutting edge analyses and advances for testing the secrets of snare is progressing. Specialists keep on investigating better approaches to test the crucial standards of quantum mechanics and to outfit the one of a kind properties of caught particles for reasonable purposes.

Trap, with its bizarre and non-instinctive properties, challenges our old style instincts and prompts significant inquiries concerning the idea of the real world. Its suggestions, both philosophical and commonsense, have upset how we might interpret the quantum world and have prompted the improvement of groundbreaking advancements, for example, quantum registering, quantum cryptography, and quantum detecting.

Trap welcomes us to rethink our traditional ideas of separateness, region, causality, and the job of awareness in molding the universe. It challenges the reductionist perspective on the real world and proposes the chance of an all encompassing, interconnected universe. The continuous discussion over the understanding of trap mirrors the profundity of the secret and the significant inquiries it raises about the idea of the universe.

As we keep on unwinding the puzzler of ensnarement and its suggestions, we adventure into an unknown area, investigating the more profound layers of the real world and the interconnectedness, everything being equal. In doing as such, we open new roads for logical revelation and mechanical progression, while likewise looking into the significant and secretive nature of the actual universe. Trap, with its "creepy activity a good ways off," keeps on being a wellspring of miracle, challenge, and motivation for researchers, logicians, and masterminds the same, welcoming us to test the limits of our comprehension and investigate the secrets of the quantum world.

Chapter 4

The Uncertainty Principle and Indeterminacy

In the domain of quantum material science, there exists a central idea that has tested our old style comprehension of the actual world - the Vulnerability Guideline. This rule, planned by the prestigious physicist Werner Heisenberg in 1927, states that it is difficult to know both the specific position and force of a subatomic molecule all the while. This idea generally changes our impression of the deterministic universe that old style physical science had long stuck to.

Heisenberg's Vulnerability Rule emerges from the wave-molecule duality of quantum mechanics. In this duality, particles show both wave-like and molecule like ways of behaving, contingent upon the conditions of perception. This implies that a molecule's situation and energy can't not entirely set in stone simultaneously, as the actual demonstration of estimating one of these properties influences the other. This indeterminacy challenges the old style idea of a precision universe in which, given adequate data, one could foresee the future with complete conviction.

The Vulnerability Guideline has expansive ramifications in how we might interpret the actual world. It isn't just a theoretical idea bound to the universe of quantum mechanics yet a central part of reality that influences our day to day routines, the improvement of innovation, and the actual texture of our universe.

One of the vital outcomes of the Vulnerability Guideline is that it brings an intrinsic unconventionality into the way of behaving of subatomic particles. This capriciousness has significant ramifications for the idea of reality itself. It actually intends that at the most essential level, the universe isn't deterministic. Regardless of whether we had total information on the underlying states of a framework, we would never foresee its future state with sureness because of the innate indeterminacy presented by the Vulnerability Guideline.

This indeterminacy challenges our natural comprehension of causality. In traditional material science, we frequently consider circumstances and logical results deterministic and unsurprising. On the off chance that we know every one of the pertinent elements, we can foresee the result of an actual interaction. In any case, the

Vulnerability Guideline recommends that there is a central cutoff to our capacity to foresee the way of behaving of the littlest particles known to mankind. This presents a degree of irregularity and capriciousness into the actual texture of the real world.

The Vulnerability Rule additionally has down to earth results in the improvement of innovation. For instance, in the field of quantum figuring, the Vulnerability Guideline forces limits on the accuracy of estimations and calculations. Quantum PCs depend on the control of quantum bits or qubits, which can exist in numerous states all the while. Notwithstanding, the actual demonstration of estimating a qubit can upset its state because of the Vulnerability Standard, making exact computations testing. This has prompted the advancement of quantum mistake remedy procedures to alleviate the impacts of indeterminacy in quantum registering.

Besides, the Vulnerability Standard assumes an essential part in quantum mechanics, impacting the way of behaving of particles on a principal level. It is the reason for the idea of wave capabilities, which portray the likelihood circulation of a molecule's situation and energy. These wave capabilities mirror the intrinsic vulnerability in these properties and are vital to how we might interpret quantum frameworks.

The Vulnerability Rule likewise reveals insight into the idea of perception and estimation in quantum mechanics. It lets us know that the demonstration of estimation itself can change the properties of a quantum framework. At the point when we measure a molecule's situation, for instance, we definitely influence its energy, as well as the other way around. This isn't an impediment of our innovation yet a basic property of the quantum world. It suggests that our perceptions are not uninvolved yet dynamic collaborations with the frameworks we study.

The philosophical ramifications of the Vulnerability Guideline are significant. It challenges our traditional thoughts of determinism and causality. It recommends that the universe is innately unsure and that there are cutoff points to our capacity to anticipate what's to come. This indeterminacy brings up issues about choice and the idea of the real world. Assuming the way of behaving of subatomic particles is innately questionable, what's the significance here for the consistency of perceptible occasions and human decisions?

A few understandings of quantum mechanics, like the Copenhagen translation, recommend that the Vulnerability Standard mirrors a crucial element of the universe. In this view, the truth is innately probabilistic, and the demonstration of estimation falls the wave capability into a specific state, bringing haphazardness into the world. Different translations, similar to the Many-Universes understanding, suggest that all potential results really happen in isolated parts of the universe, taking out the requirement for breakdown and safeguarding determinism in a multiverse of potential outcomes.

The Vulnerability Guideline has likewise prompted banters about the idea of the real world and the job of the spectator. Some contend that the demonstration of perception is essential to the way of behaving of quantum frameworks, while others battle that there should be a fundamental, objective reality that exists autonomously of perception. These philosophical discussions feature the significant effect of the Vulnerability Standard on how we might interpret the idea of the universe.

Notwithstanding its ramifications for quantum mechanics and reasoning, the Vulnerability Rule has pragmatic applications in different fields. It plays had a critical impact in the improvement of current innovation, for example, the creation of the electron magnifying lens, which utilizes the Vulnerability Standard to accomplish very high-goal imaging of nanoscale objects. Additionally, in atomic material science, the Vulnerability Rule is fundamental for figuring out the way of behaving of nuclear cores and the cycles of atomic responses.

The Vulnerability Rule has likewise tracked down applications in fields past physical science, like craftsmanship and writing. A few craftsmen and journalists have drawn motivation from the possibility of indeterminacy and haphazardness in the universe. They investigate the idea of vulnerability in their imaginative works, pondering its philosophical and existential ramifications.

In rundown, the Vulnerability Guideline, planned by Werner Heisenberg, is a fundamental idea in quantum material science that challenges our old style comprehension of the deterministic universe. It brings inborn indeterminacy into the way of behaving of subatomic particles, influencing how we might interpret causality, estimation, and the idea of the real world. This rule has significant ramifications for innovation, reasoning, and the manner in which we see the world. It is an update that, at the most basic level, the universe stays questionable, puzzling, and loaded with shocks.

4.1 Heisenberg's Uncertainty Principle and its significance

Werner Heisenberg's Vulnerability Guideline remains as perhaps of the most significant and persevering through idea in the domain of physical science. Formed in 1927, this guideline profoundly modified how we might interpret the subatomic world and presented a degree of capriciousness and indeterminacy that tested the determinism of traditional material science. In this investigation, we dig into Heisenberg's Vulnerability Guideline and its sweeping importance, both inside the domain of quantum physical science and then some.

At the core of the Vulnerability Guideline is the idea that it is difficult to know both the specific position and force of a subatomic molecule at the same time. This apparently straightforward assertion conveys significant ramifications for our perception of the actual universe. It emerges from the wave-molecule duality innate in quantum mechanics, a duality that recommends particles can show both wave-like and molecule like ways of behaving. Thus, any endeavor to exactly decide both the position and force of a molecule turns out to be intrinsically restricted, as the demonstration of estimating one property definitely influences the other.

The ramifications of the Vulnerability Rule stretch out a long ways past the domain of hypothetical physical science; they have significant importance for how we might interpret the idea of the real world, the improvement of innovation, and, surprisingly, philosophical and existential inquiries.

One of the most prompt results of the Vulnerability Guideline is the presentation of intrinsic eccentricism into the way of behaving of subatomic particles. This unusualness comes from the way that, at the quantum level, the universe isn't deterministic. Regardless of whether one were equipped with complete information on the underlying states of a framework, Heisenberg's guideline forces a crucial breaking point on the capacity to foresee the framework's future state with sureness.

Generally, the Vulnerability Guideline challenges the traditional idea of causality as a deterministic and unsurprising idea. In old style material science, one frequently considers causality a chain of occasions that can be perceived and anticipated unhesitatingly assuming one has adequate data. Nonetheless, the Vulnerability Guideline affirms that there is an innate cutoff to our prescient powers, because of the characteristic indeterminacy presented at the quantum level. This key degree of flightiness brings a component of haphazardness into the universe, an idea that profoundly modifies how we might interpret the idea of the real world.

The Vulnerability Guideline's useful outcomes are especially clear in the advancement of innovation. Quantum processing, for example, is a field where the Vulnerability Standard forces significant cutoff points on accuracy. Quantum PCs depend on controlling quantum bits or qubits, which can exist in numerous states at the same time, on account of the superposition rule. In any case, the actual demonstration of estimating a qubit, which is important for getting data, can upset its state because of the Vulnerability Standard. This aggravation makes exact estimations testing and requires the improvement of quantum blunder adjustment methods to moderate the impacts of indeterminacy in quantum figuring.

Moreover, the Vulnerability Guideline assumes a significant part in the hypothesis of quantum mechanics itself. It impacts the way of behaving of particles on a principal level and is vital to how we might interpret quantum frameworks. In quantum mechanics, the idea of a wave capability emerges, portraying the likelihood conveyance of a molecule's situation and energy.

These wave capabilities mirror the characteristic vulnerability in these properties and support the numerical structure that depicts quantum peculiarities.

The Vulnerability Standard likewise develops how we might interpret the course of perception and estimation in quantum mechanics. It enlightens the way that the actual demonstration of estimation influences the properties of a quantum framework. At the point when one estimates a molecule's situation, for example, it inescapably impacts its energy, as well as the other way around. This isn't a restriction of our innovation or observational abilities yet an inborn element of the quantum world. It proposes that our perceptions are not aloof demonstrations of just uncovering prior properties however dynamic connections with the frameworks we study.

The philosophical ramifications of the Vulnerability Rule are both significant and persevering. This guideline challenges old style ideas of determinism and causality. It recommends that the universe is innately dubious, and there are inborn cutoff points to our capacity to foresee what's in store. These ramifications bring up issues about freedom of thought and the actual idea of the real world. Assuming that the way of behaving of subatomic particles is essentially unsure, what's the significance here for the consistency of plainly visible occasions and human decisions?

Translations of quantum mechanics, like the Copenhagen understanding, suggest that the Vulnerability Guideline is a major element of the universe. In this view, the truth is innately probabilistic, and the demonstration of estimation falls the wave capability into a specific state, bringing irregularity into the world. Different translations, similar to the Many-Universes understanding, recommend that all potential results really happen in isolated parts of the universe, in this way disposing of the requirement for wave capability breakdown and safeguarding determinism in a multiverse of conceivable outcomes.

The Vulnerability Rule likewise ignites banters about the idea of the real world and the job of the spectator. Some contend that the demonstration of perception is major to the way of behaving of quantum frameworks, recommending that reality itself might rely upon cognizant perception. Others declare that there should be a fundamental, objective reality that exists autonomously of perception, and the job of the eyewitness is more inactive, only uncovering the properties that were at that point there.

Notwithstanding its ramifications for quantum mechanics and reasoning, the Vulnerability Rule tracks down pragmatic applications in different fields. It plays had a crucial impact in the improvement of current innovation. For example, the creation of the electron magnifying instrument is a demonstration of the Vulnerability Guideline's importance. This strong imaging device depends on Heisenberg's guideline to accomplish uncommonly high-goal imaging of nanoscale objects. In the domain of atomic material science, the Vulnerability Guideline is fundamental for grasping the way of behaving of nuclear cores and the cycles of atomic responses.

In addition, the Vulnerability Standard has made a permanent imprint on fields past physical science. Specialists and authors have tracked down motivation in the idea of indeterminacy and arbitrariness. They investigate the possibility of vulnerability in their imaginative works, considering its philosophical and existential ramifications. This transaction among workmanship and science features the significant and interdisciplinary nature of Heisenberg's Vulnerability Guideline.

All in all, Heisenberg's Vulnerability Rule, made by Werner Heisenberg in 1927, is a foundation of quantum physical science that challenges our traditional comprehension of a deterministic universe. It brings inherent indeterminacy into the way of behaving of subatomic particles, reshaping our perception of causality, estimation, and the actual texture of the real world. This standard holds significant ramifications for innovation, theory, and our more extensive perspective. It fills in as an update that, at the most basic level, the universe stays questionable, confounding, and brimming with secrets, welcoming us to investigate the profundities of quantum reality and its persevering through importance.

4.2 The limits of precision in quantum measurements

Quantum mechanics, with its wave-molecule duality and innate vulnerability, challenges our old style instincts about the actual world. One of the central rules that exemplifies this challenge is Heisenberg's Vulnerability Standard, which states that it is difficult to know both the specific position and energy of a subatomic molecule all the while. This rule has significant ramifications for the accuracy of quantum estimations and the limits innate in our endeavors to figure out the quantum domain.

At its center, the Vulnerability Standard emerges from the wave-molecule duality of quantum particles. This duality implies that particles can display both wave-like and molecule like ways of behaving, contingent upon the setting of perception. At the point when particles are treated as waves, their positions are portrayed by likelihood disseminations, and their momenta are associated with the properties of these wave capabilities. This wave-molecule duality acquaints a principal limit with the accuracy with which we can know both position and force, as the demonstration of estimating one property definitely upsets the other.

The Vulnerability Rule can be numerically communicated as $\Delta x \delta p \geq \hbar/2$, where Δx addresses the vulnerability ready, Δp addresses the vulnerability in force, and $\hbar$ (h-bar) is the diminished Planck steady, a principal consistent of nature. This disparity lets us know that the result of the vulnerabilities ready and energy should be more noteworthy than or equivalent to $\hbar/2$. In useful terms, the more exactly we know a molecule's situation (Δx diminishes), the less definitively we can know its force (Δp increments), as well as the other way around.

The Vulnerability Rule has a few ramifications for the accuracy of quantum estimations. It, first and foremost, intends that there is a central cutoff to how exactly we can all the while know position and force. This breaking point isn't a consequence of constraints in our estimation innovation however is characteristic for the idea of the quantum world.

Regardless of how cutting-edge our instruments become, we can never defeat this central limitation forced by the Vulnerability Guideline.

Also, the Vulnerability Rule suggests that the accuracy of our estimations is restricted by the size of $\hbar$, which is a tiny worth. The decreased Planck steady, $\hbar$, is around $1.0545718 \times 10^{-34}$ Joule-seconds, and this little worth sets the scale for the vulnerability in quantum estimations. This implies that while managing perceptible items, the vulnerability ready and force is immaterial and frequently vague in our regular encounters. Be that as it may, at the subatomic level, where particles are a lot more modest and their properties are more questionable, the impacts of the Vulnerability Rule become critical.

To more readily comprehend the restrictions forced by the Vulnerability Standard, we should think about two key situations: position estimations and energy estimations.

Position Estimations:

With regards to situate estimations, the Vulnerability Standard lets us know that the more exactly we endeavor to gauge the place of a molecule, the less precisely we can decide its force. As such, if we need to find a molecule with high accuracy, we should acknowledge that our insight into its force will be somewhat questionable. This relationship generally restricts the accuracy of gadgets like electron magnifying lens, which depend on the place of electrons to deliver high-goal pictures.

For instance, think about an electron in circle around a core in a molecule. In the event that we endeavor to decide its situation with very high accuracy, we should utilize a short-frequency test like X-beams or high-energy electrons. These short-frequency tests can pinpoint the electron's position all the more precisely, however the related energy is adequately high to altogether upset the electron's force. Subsequently, the electron's energy turns out to be profoundly dubious, and our synchronous information on its situation and force becomes restricted by the Vulnerability Guideline.

Force Estimations:

On the other hand, when we center around energy estimations, the Vulnerability Rule directs that the more exactly we endeavor to gauge the force of a molecule, the less precisely we can decide its situation. By and by, this truly intends that to know a molecule's energy with high accuracy, we should acknowledge that our insight into its position will be somewhat questionable. The constraints forced by the Vulnerability Standard become obvious when we see gadgets like molecule gas pedals, which expect to decide the momenta of subatomic particles with extraordinary accuracy.

In an atom smasher, like the Huge Hadron Collider (LHC), scientists speed up particles to exceptionally high velocities and impact them to concentrate on their properties. To decide a molecule's force with high accuracy, one should grant a lot of energy to the molecule.

In any case, this high-energy collaboration unavoidably brings about a significant vulnerability in the molecule's situation. Fundamentally, the Vulnerability Rule limits the accuracy with which we can all the while know both position and force.

These models outline the central compromise among position and energy estimations directed by the Vulnerability Guideline. The more we mean to upgrade accuracy in one angle, the more vulnerability is brought into the other. It is fundamental to perceive that these restrictions are not a consequence of our estimation devices or strategies; they are inborn to the quantum world and are key parts of the universe.

The Vulnerability Standard's importance broadens well past its part in quantum mechanics and the accuracy of estimations. It has significant ramifications for how we might interpret causality, determinism, and the idea of reality itself.

One of the vital philosophical ramifications of the Vulnerability Rule is its test to the traditional thought of causality. In old style physical science, we frequently consider circumstances and logical results deterministic and unsurprising. In the event that we have total information on the underlying states of a framework, we can hypothetically foresee its future state with conviction. Nonetheless, the Vulnerability Guideline recommends that there are principal cutoff points to our prescient powers. It presents a component of irregularity and flightiness at the quantum level, testing the possibility of a precision universe in which all occasions are foreordained.

The Vulnerability Rule brings up issues about the idea of determinism and freedom of thought. Assuming that the way of behaving of subatomic particles is essentially questionable, what does this infer for the consistency of naturally visible occasions and human decisions? A few translations of quantum mechanics, similar to the Copenhagen understanding, suggest that the Vulnerability Rule mirrors a basic indeterminacy in the universe. In this view, the truth is intrinsically probabilistic, and the demonstration of estimation falls the wave capability into a specific state, bringing haphazardness into the world. Different understandings, like the Many-Universes translation, recommend that all potential results really happen in discrete parts of the universe, saving determinism in a multiverse of conceivable outcomes.

The Vulnerability Rule additionally has reasonable ramifications for innovation and logical exploration. In quantum figuring, for instance, the Vulnerability Standard forces limits on the accuracy of estimations and calculations. Quantum PCs depend on the control of quantum bits or qubits, which can exist in various states at the same time because of the rule of superposition. Be that as it may, the actual demonstration of estimating a qubit can upset its state because of the Vulnerability Standard, making exact computations testing. This challenge has prompted the advancement of quantum blunder amendment strategies to alleviate the impacts of indeterminacy in quantum figuring.

Besides, the Vulnerability Rule significantly affects how we might interpret the way of behaving of particles in the quantum world. It is the establishment for the idea of wave capabilities, which depict the likelihood dissemination of a molecule's situation and energy. These wave capabilities mirror the inborn vulnerability in these properties and are fundamental to our perception of quantum frameworks.

The Vulnerability Rule likewise enlightens the job of perception and estimation in quantum mechanics. It lets us know that the demonstration of estimation itself can modify the properties of a quantum framework. At the point when we measure a molecule's situation, for example, we inescapably impact its energy, as well as the other way around. This isn't a limit of our innovation however a key property of the quantum world, underlining that our perceptions are not inactive yet dynamic communications with the frameworks we study.

Notwithstanding its ramifications for quantum mechanics and innovation, the Vulnerability Standard tracks down applications in different fields, including workmanship and writing. A few specialists and essayists draw motivation from the possibility of indeterminacy and haphazardness in the universe. They investigate the idea of vulnerability in their imaginative works, considering its philosophical and existential ramifications. This interdisciplinary interchange among science and artistic expressions highlights the significant and general nature of Heisenberg's Vulnerability Rule.

The ramifications of the Vulnerability Rule have prompted philosophical discussions about the idea of the real world and the job of the eyewitness. Some contend that the demonstration of perception is basic to the way of behaving of quantum frameworks, proposing that reality itself might rely upon cognizant perception. This point of view has led to thoughts like the onlooker impact, which places that the demonstration of perception can impact the result of quantum tests. Others affirm that there should be a basic, objective reality that exists freely of perception, and the job of the eyewitness .

4.3 The philosophical implications of indeterminacy

The idea of indeterminacy, as uncovered by Heisenberg's Vulnerability Guideline in the domain of quantum mechanics, has significant philosophical ramifications that stretch out past the limits of physical science. At its center, indeterminacy challenges old style ideas of determinism, causality, and the idea of reality itself. In this investigation, we dig into the philosophical ramifications of indeterminacy and how it has molded how we might interpret the universe, human organization, and the actual texture of presence.

The Vulnerability Rule, as formed by Werner Heisenberg in 1927, states that it is difficult to know both the specific position and force of a subatomic molecule all the while. This guideline emerges from the wave-molecule duality of quantum mechanics, which recommends that particles can display both wave-like and molecule like ways of behaving, contingent upon the setting of perception.

At the point when particles are treated as waves, their positions are portrayed by likelihood disseminations, and their momenta are associated with the properties of these wave capabilities. Subsequently, any endeavor to exactly decide both the position and force of a molecule turns out to be intrinsically restricted, as the demonstration of estimating one property unavoidably upsets the other.

The Vulnerability Standard can be numerically communicated as $\Delta x \delta p \geq \hbar/2$, where Δx addresses the vulnerability ready, Δp addresses the vulnerability in force, and $\hbar$ (h-bar) is the diminished Planck steady. This imbalance embodies the significant ramifications of indeterminacy for how we might interpret the quantum world.

The first and most prompt philosophical ramifications of indeterminacy is its test to the old style thought of causality. In traditional physical science, the universe is many times seen as a deterministic and unsurprising framework. In the event that we have total information on the underlying states of a framework, we can hypothetically foresee its future state unhesitatingly. This deterministic perspective on the universe, frequently alluded to as the "precision universe," expects that all occasions are destined by going before occasions in a constant chain of circumstances and logical results.

Nonetheless, indeterminacy, as uncovered by the Vulnerability Guideline, presents an inborn component of arbitrariness and capriciousness at the quantum level. That's what it proposes, regardless of whether we had total information on the underlying states of a quantum framework, there are crucial cutoff points to our prescient powers. The demonstration of estimation, which unavoidably upsets a quantum framework, presents a component of indeterminacy and irregularity that challenges the deterministic perspective on the universe. This brings up a major issue: Could we at any point really foresee the future with sureness, or is the universe intrinsically questionable?

The Vulnerability Rule additionally has huge ramifications for the idea of the real world. It prompts us to reexamine the old style thought of a goal, onlooker autonomous reality. In traditional physical science, it is in many cases accepted that reality exists autonomously of perception. Onlookers simply reveal the previous properties of articles through estimations. In this view, the job of the onlooker is aloof, and the properties of the noticed article are viewed as determinate and fixed.

Nonetheless, the Vulnerability Standard presents a significant change in context. It lets us know that the demonstration of estimation is definitely not a uninvolved cycle however a functioning connection with the quantum framework. At the point when we measure a molecule's situation, for instance, we inescapably impact its energy, as well as the other way around. This proposes that the onlooker is a basic piece of the quantum framework and that the properties of the noticed article are not fixed freely of perception. All things considered, they are dependent upon the demonstration of estimation.

This change in context challenges our old style ideas of an objective reality. It proposes that reality might be more subject to the demonstration of perception than recently suspected. This point of view has prompted banters about the job of cognizance and the onlooker in forming the quantum world. A few understandings of quantum mechanics, like the Copenhagen translation, recommend that the demonstration of estimation falls the wave capability into a specific state, and reality becomes dependent upon cognizant perception. In this view, cognizance assumes a crucial part in molding the quantum world.

Alternately, different translations, similar to the Many-Universes understanding, propose an alternate answer for the issue of indeterminacy. As per this translation, all potential results of a quantum estimation really happen in isolated parts of the universe. In this multiverse of potential outcomes, determinism is protected, and the job of the spectator is to follow one specific part of the real world, while any remaining prospects unfurl in equal. This translation dispenses with the requirement for wave capability breakdown and keeps up with the thought of an objective reality, but in an immeasurably extended multiverse.

The Vulnerability Standard and indeterminacy additionally bring up issues about human office and choice. Assuming the way of behaving of subatomic particles is intrinsically unsure and impacted by the demonstration of perception, what's the significance here for the consistency of naturally visible occasions and human decisions? It challenges the traditional thought that human activities and still up in the air by going before occasions and can be anticipated with assurance assuming we have total information on the underlying circumstances.

A few savants and researchers contend that the inborn indeterminacy at the quantum level doesn't be guaranteed to infer an indeterminacy at the plainly visible level. They recommend that the probabilistic idea of quantum occasions doesn't be guaranteed to convert into an absence of determinism in regular day to day existence. In any case, the specific connection between quantum indeterminacy and human organization stays a subject of continuous discussion.

The Vulnerability Guideline additionally welcomes us to reevaluate the idea of logical information and its restrictions. In old style physical science, the objective of science is frequently viewed as revealing goal and certain insights about the world. Logical regulations and hypotheses are viewed as precise portrayals of a hidden, determinate reality. Nonetheless, the Vulnerability Guideline advises us that our logical information is obliged by the inborn impediments of the quantum world.

In quantum mechanics, our insight is described by probabilities and vulnerabilities. We can make forecasts about the probability of different results, yet we can't make deterministic expectations with full confidence. This brings up issues about the idea of logical information and the degree to which it can give evenhanded and certain bits of insight about the world.

Besides, the Vulnerability Rule has down to earth suggestions for innovation and logical exploration. It provokes our capacity to make exact estimations and computations, especially in the improvement of trend setting innovations, for example, quantum processing. Quantum PCs depend on the control of quantum bits or qubits, which can exist in numerous states at the same time because of the rule of superposition. Nonetheless, the actual demonstration of estimating a qubit can upset its state because of the Vulnerability Rule, making exact computations testing. This challenge has prompted the improvement of quantum blunder revision procedures to alleviate the impacts of indeterminacy in quantum figuring.

Notwithstanding its ramifications for quantum mechanics and reasoning, the Vulnerability Rule tracks down applications in different fields, including workmanship and writing. A few craftsmen and scholars draw motivation from the possibility of indeterminacy and irregularity in the universe. They investigate the idea of vulnerability in their imaginative works, considering its philosophical and existential ramifications. This interdisciplinary interaction among science and human expressions highlights the significant and general nature of the Vulnerability Guideline.

In rundown, the philosophical ramifications of indeterminacy, as uncovered by Heisenberg's Vulnerability Standard, challenge traditional thoughts of determinism, causality, and the idea of the real world. This rule prompts us to reevaluate the consistency of the universe, the job of the spectator in forming reality, and the connection between quantum indeterminacy and human organization. It likewise brings up issues about the idea of logical information and its limits. The Vulnerability Rule welcomes us to investigate the profundities of quantum reality and its effect on how we might interpret the world, empowering us to embrace the inborn vulnerability and intricacy of the universe.

Chapter 5

Quantum Mechanics and the Nature of Space and Time

The investigation of quantum mechanics changed how we might interpret the actual world, testing laid out old style hypotheses and offering new bits of knowledge into the idea of reality. Quantum mechanics, formed in the mid twentieth hundred years, presented a change in perspective that significantly affected our view of the texture of the universe.

Key to the ramifications of quantum mechanics is the wave-molecule duality, a major idea that re-imagined how we might interpret matter. This duality recommends that particles, like electrons and photons, can show both wave-like and molecule like ways of behaving relying upon the trial conditions. At the core of this peculiarity lies the vulnerability guideline, broadly planned by Werner Heisenberg in 1927. This guideline states that the position and force of a molecule can't be at the same time estimated with outright accuracy. The more exactly one amount is known, the more dubious the estimation of the other becomes.

The wave-molecule duality and the vulnerability standard tested the deterministic perspective of traditional physical science, featuring the intrinsic eccentricism and indeterminacy at the quantum level. These ideas have broad ramifications for how we might interpret existence, bringing up significant issues about the idea of the real world and the essential construction of the universe.

One of the key difficulties quantum mechanics presents is to our traditional comprehension of existence. In old style physical science, existence were viewed as particular and outright, shaping the stage on which all actual occasions happened. Time was remembered to stream consistently and freely of occasions, and space was viewed as a fixed and constant territory in which matter and energy cooperated.

Nonetheless, quantum mechanics presents a degree of vulnerability and indeterminacy that challenges this old style origination. The vulnerability guideline essentially adjusts the manner in which we see the connection among reality at the quantum level. It suggests that there are intrinsic restrictions to the accuracy with which we can gauge both the position and force of particles, upsetting the old style thought of a distinct, ceaseless reality.

Quantum mechanics brings up issues about the basic granularity of reality. At the quantum level, the actual idea of a consistent, fixed reality is raised doubt about. A few hypotheses in quantum gravity, like circle quantum gravity and string hypothesis, recommend that existence could have a discrete, granular design at tiny scopes, testing the smooth and persistent nature of traditional spacetime.

These speculations suggest that at the Planck scale, which is a staggeringly limited scale of around 10^{-35} meters, spacetime may be made out of inseparable units or "quanta" of existence. These key units could lead to a design that is unfathomably not quite the same as the persistent and smooth spacetime of old style physical science. This granular construction, whenever affirmed, would have critical ramifications for how we might interpret the idea of reality and could reform our originations of the texture of the universe.

The vulnerability rule additionally difficulties how we might interpret the restriction of particles in reality. In traditional material science, particles were remembered to have obvious positions and momenta. Notwithstanding, in the quantum domain, the actual demonstration of estimation influences the position and force of particles, prompting an inborn vulnerability in the two amounts.

This difficulties the traditional thought of a flat out and unmistakable place of a molecule in space. Quantum mechanics infers that particles don't have an exact area until they are estimated. Until the estimation happens, a molecule exists in a superposition of potential expresses, a condition of probability addressed by a wave capability. This superposition proposes that a molecule might exist in various places at the same time until it is noticed, obscuring the customary thoughts of spatial limitation.

Besides, quantum snare, a peculiarity where particles become interconnected no matter what the distance isolating them, challenges how we might interpret the freedom of existence.

Trapped particles appear to divide an association that rises above the spatial detachment among them. Changes in a single molecule's state promptly influence the other, apparently disregarding the requirements of traditional ideas of region and distinctness.

This non-region alludes to a more profound interconnectedness in the texture of reality than traditional material science represented. It brings up issues about the fundamental idea of room time connections and the potential interconnectedness that might exist past the old style limits of reality.

Quantum mechanics likewise represents a test to how we might interpret the bolt of time, which alludes to the unevenness of time from past to future. In old style physical science, time was accepted to advance consistently from the past to the future, following a deterministic way characterized by the laws of circumstances and logical results.

In any case, quantum mechanics presents a period lopsidedness at the quantum level that isn't obviously reconcilable with the old style thought of time's unidirectional stream. In quantum frameworks, processes are reversible, implying that the development of a framework from a past state to a future not entirely set in stone by the laws of physical science alone. All things being equal, the laws of quantum mechanics consider processes that are similarly conceivable in one or the other heading of time.

This difficulties our traditional comprehension of time's unidirectional stream and the deterministic walk of occasions from the past to what's in store. The reversible idea of quantum processes brings up issues about the crucial imbalance of time and whether time's bolt may be an emanant property at bigger scopes, instead of a key part of the quantum world.

Moreover, the idea of superposition in quantum mechanics challenges how we might interpret the progression of time. The possibility that particles can exist in numerous states all the while until noticed obscures the customary ideas of the consecutive entry of time. In the quantum domain, a molecule's state doesn't implode until estimated, suggesting that different potential states coincide until noticed.

This difficulties our instinctive comprehension of the progression of time as a successive movement of occasions starting with one state then onto the next. All things considered, quantum mechanics recommends a non-straight point of view where different potential states exist simultaneously until perception happens, welcoming a reconsideration of the consecutive progression of time.

The philosophical ramifications of quantum mechanics for the idea of existence stretch out to how we might interpret causality and determinism. In traditional material science, causality was viewed as a crucial rule where the current situation with the not entirely settled by its past states.

Situation transpired in a deterministic way, and what was in store was accepted to be very much unsurprising in the event that one had total information on the underlying circumstances.

In any case, the intrinsic indeterminacy and flightiness presented by the vulnerability rule challenge the deterministic perspective. Quantum mechanics recommends that at the quantum level, there are crucial cutoff points to our capacity to anticipate what's in store. The demonstration of estimation presents a component of irregularity and unusualness, which challenges the old style idea of a perfect timing universe with foreordained occasions.

This presents inquiries concerning the job of possibility and freedom of thought in the universe. Assuming the way of behaving of particles is innately questionable and affected by perception, what's the significance here for the consistency of naturally visible occasions and human decisions? The vulnerability guideline brings up issues about the limits of determinism and the possible job of chance known to mankind.

The connection between quantum mechanics and the idea of existence likewise reaches out to the investigation of dark openings and the idea of singularities. Dark openings are districts of room where gravitational powers are extreme to such an extent that nothing, not even light, can get away from their force. At the core of a dark opening falsehoods a peculiarity, a place of boundless thickness where the laws of material science separate.

The investigation of dark openings and singularities brings up issues about the constraints of how we might interpret the texture of existence. Quantum mechanics, related to general relativity, has incited the quest for a hypothesis of quantum gravity, which intends to accommodate the laws of quantum mechanics with the hypothesis of general relativity to figure out the way of behaving of issue and energy at outrageous scales.

5.1 Quantum mechanics and its influence on our perception of space and time

Quantum mechanics, the part of physical science that arrangements with the way of behaving of particles on the littlest scales, has been quite possibly of the most progressive and perplexing hypothesis throughout the entire existence of science. It tested traditional physical science as well as reshaped our impression of reality. At the core of this groundbreaking worldview lies the wave-molecule duality and Heisenberg's Vulnerability Standard, which brought significant bits of knowledge and intricacies into the idea of the real world, space, and time.

Quantum mechanics in a general sense modified how we might interpret the idea of issue. One of the focal fundamentals of quantum mechanics is the wave-molecule duality. This idea affirms that particles, like electrons and photons, can show both wave-like and molecule like ways of behaving relying upon the setting of perception. Basically, it obscures the lines between particles as limited substances and waves as conveyed peculiarities, provoking a crucial change in how we might interpret matter.

The wave-molecule duality is best exemplified by the renowned twofold cut try. In this trial, particles, like electrons, are sent each in turn through two cuts. Shockingly, when the electrons are not noticed, they make an impedance design on the screen behind the cuts, looking like the example framed by waves. This suggests that the electrons are acting like waves, displaying wave-like impedance.

In any case, when the electrons are noticed or estimated, they act as discrete particles, going through one of the cuts and not making an obstruction design. This analysis exhibits that the actual demonstration of perception impacts the way of behaving of particles, prompting the conjunction of both wave and molecule attributes.

This duality brings up significant issues about the idea of particles, space, and time. It infers that particles don't have distinct positions and momenta until they are estimated. All things considered, they exist in a superposition of likely states until perception falls their wave capability into a specific state. This difficulties our traditional thoughts of a persistent and deterministic reality and hazy spots the limits between the material and the insignificant.

Heisenberg's Vulnerability Standard, figured out by Werner Heisenberg in 1927, is one more foundation of quantum mechanics with critical ramifications for how we might interpret existence. This rule states that there is a characteristic breaking point to the accuracy with which we can at the same time know both the position and force of a molecule. The more exactly one not set in stone, the more unsure the estimation of the other becomes.

Numerically, the Vulnerability Guideline is communicated as $\Delta x \delta p \geq \hbar/2$, where Δx addresses the vulnerability ready, Δp addresses the vulnerability in energy, and $\hbar$ (h-bar) is the decreased Planck consistent. This crucial disparity underlines the innate impediments of our insight about the condition of particles at the quantum level.

The Vulnerability Standard difficulties our old style comprehension of the restriction of particles in existence. In traditional material science, particles were remembered to have distinct positions and momenta, considering exact forecasts about their future way of behaving. Nonetheless, the Vulnerability Standard infers that there are essential cutoff points to our capacity to know both the position and force of particles exactly. This presents an innate indeterminacy in how we might interpret the quantum world, upsetting the old style thought of a nonstop, clear cut reality.

The ramifications of quantum mechanics for our impression of existence are broad. They challenge the traditional thoughts of a deterministic and persistent universe and open up new philosophical inquiries regarding the idea of the real world.

One of the critical philosophical ramifications of quantum mechanics is its test to the traditional idea of causality. In old style physical science, causality was perceived as a deterministic and unsurprising chain of occasions, where the current situation with the not entirely settled by its past states. Situation transpired in a circumstances and logical results way, and what was in store was viewed as no doubt unsurprising if one had total information on the underlying circumstances.

In any case, quantum mechanics presents a component of arbitrariness and unusualness at the quantum level, as shown by the Vulnerability Rule. This difficulties the deterministic perspective of old style material science. That's what it proposes, regardless of whether we had total information on the underlying states of a quantum framework, there are essential cutoff points to our prescient powers. The demonstration of estimation presents a component of indeterminacy and haphazardness that disturbs the deterministic perspective on the universe, bringing up issues about the consistency of naturally visible occasions and the job of chance on the planet.

The idea of wave capabilities in quantum mechanics further convolutes how we might interpret causality. Wave capabilities portray the likelihood dispersion of a molecule's situation and force and mirror the intrinsic vulnerability in these properties. These wave capabilities address a condition of probability, proposing that particles exist in numerous states all the while until noticed. This difficulties the traditional thought of a direct and deterministic causality, where situation unfurl from past to future in a successive and unsurprising way. All things considered, quantum mechanics suggests a more intricate interaction between likely states and the impact of perception on the course of occasions.

Quantum mechanics likewise challenges our old style originations of the idea of reality. In old style material science, reality were viewed as unmistakable and outright, shaping the setting against which all actual occasions happened. Time was remembered to stream consistently and freely of occasions, while space was viewed as a fixed and ceaseless field in which matter and energy collaborated.

In any case, quantum mechanics presents a degree of vulnerability and indeterminacy that disturbs this old style origination. The Vulnerability Guideline in a general sense changes the manner in which we see the connection among reality at the quantum level. It infers that there are innate limits to the accuracy with which we can gauge both the position and force of particles, testing the old style thought of an obvious, nonstop reality.

Quantum mechanics brings up issues about the basic granularity of existence. A few speculations in quantum gravity, like circle quantum gravity and string hypothesis, recommend that existence could have a discrete, granular construction at minuscule scopes. These hypotheses recommend that at the Planck scale, which is an extraordinarily limited scale of around 10^{-35} meters, spacetime may be made out of unified units or "quanta" of existence. This granular design difficulties the smooth and consistent spacetime of traditional material science and presents the idea of a generally discrete and quantized spacetime.

This idea of a granular spacetime prompts a reconsideration of our traditional originations of reality. It infers that at the quantum level, existence may not be consistent but instead made out of discrete units, obscuring the customary limits between the ceaseless and the discrete.

The idea of quantum trap further difficulties our old style comprehension of the autonomy of existence. Trapped particles, no matter what the distance isolating them, show a significant association where changes in a single molecule's state immediately influence the other, apparently disregarding the requirements of old style ideas of territory and distinguishableness.

This non-territory alludes to a more profound interconnectedness in the texture of existence than traditional physical science represented. It brings up issues about the basic idea of room time connections and the potential interconnectedness that might exist past the traditional limits of reality.

Quantum mechanics likewise represents a test to how we might interpret the bolt of time, which alludes to the unevenness of time from past to future. In old style physical science, time was accepted to advance consistently from the past to the future, following a deterministic way characterized by the laws of circumstances and logical results.

Nonetheless, quantum mechanics presents a period unevenness at the quantum level that isn't effectively reconcilable with the traditional thought of time's unidirectional stream. In quantum frameworks, processes are reversible, implying that the development of a framework from a past state to a future not entirely set in stone by the laws of physical science alone. All things being equal, the laws of quantum mechanics consider processes that are similarly conceivable in one or the other bearing of time.

This difficulties our old style comprehension of time's unidirectional stream and the deterministic movement of occasions from the past to what's in store. The reversible idea of quantum processes brings up issues about the key deviation of time and whether time's bolt may be a developing property at bigger scopes, instead of a central part of the quantum world.

Moreover, the idea of superposition in quantum mechanics challenges how we might interpret the progression of time. In the quantum domain, a molecule's state doesn't implode until estimated, suggesting that different potential states exist together until perception happens. This difficulties our natural comprehension of the progression of time as a consecutive movement of occasions starting with one state then onto the next. In the quantum domain, various potential states exist simultaneously until perception, welcoming a reexamination of the consecutive progression of time.

The philosophical ramifications of quantum mechanics for our view of reality stretch out to the investigation of dark openings and the idea of singularities. Dark openings are locales of room where gravitational powers are serious to the point that nothing.

5.2 Relativity and quantum physics: Bridging the gap

The domains of relativity and quantum physical science address two of the best and persuasive speculations throughout the entire existence of physical science. Created in the mid twentieth 100 years, Albert Einstein's hypothesis of relativity changed how we might interpret gravity, space, and time. At the same time, quantum mechanics, with its central standards of wave-molecule duality and the Vulnerability Rule, reformed our perception of the minute world. However, in spite of their singular triumphs, these two hypotheses appear to possess in a general sense various universes - one that depicts the naturally visible and vast sizes of the universe, and the other that disentangles the complexities of the quantum domain. Overcoming any issues between these two spaces has been perhaps of the most significant and tricky test in the field of hypothetical physical science. In this investigation, we dive into the captivating journey to accommodate relativity and quantum physical science and the continuous undertakings to make a brought together hypothesis that can envelop both the plainly visible and tiny parts of our universe.

Relativity: A Plainly visible Insurgency

Albert Einstein's hypothesis of relativity, contained two significant parts - Unique Relativity and General Relativity, has fundamentally affected our impression of the universe on plainly visible scales. Exceptional Relativity, formed in 1905, presented the well known condition $E=mc^2$, showing the proportionality of mass and energy. It additionally reclassified how we might interpret existence, especially with the idea of time expansion, which expresses that time elapses diversely for eyewitnesses in relative movement.

The hypothesis of General Relativity, created in 1915, developed the establishment laid by Exceptional Relativity. It depicted gravity as the shape of spacetime brought about by the presence of mass and energy. General Relativity richly made sense of the precession of the perihelion of Mercury, a longstanding riddle in traditional physical science, and anticipated the presence of dark openings, gravitational waves, and the twisting of light in the gravitational field of huge items.

Notwithstanding, General Relativity had its own arrangement of difficulties. While it effectively portrayed the gravitational way of behaving of huge scope objects and inestimable peculiarities, it was generally inconsistent with quantum mechanics, the hypothesis that oversees the way of behaving of particles at the subatomic level.

Quantum Mechanics: A Minute Unrest

Quantum mechanics, which arose around a similar time as Einstein's hypothesis of relativity, achieved an upset in how we might interpret the minute world. It presented the idea of wave-molecule duality, which recommended that particles, like electrons and photons, could show both wave-like and molecule like ways of behaving.

This duality, exemplified by the twofold cut explore, showed the way that particles could disrupt themselves, making examples of impedance that suggested both molecule and wave qualities.

At the core of quantum mechanics lies the Vulnerability Guideline, figured out by Werner Heisenberg in 1927. This guideline expresses that it is difficult to know both the specific position and energy of a molecule with full confidence all the while. The more definitively one amount is estimated, the more unsure the estimation of the other becomes. This key rule brought innate indeterminacy into the way of behaving of particles at the quantum level, testing old style thoughts of determinism and consistency.

Quantum mechanics likewise uncovered the peculiarity of quantum ensnarement, where particles become interconnected so that the properties of one molecule immediately influence the properties of another, no matter what the distance isolating them. This non-nearby connection appeared to resist the traditional idea of territory and distinguishableness, further featuring the baffling idea of the quantum domain.

Quantum mechanics, with its probabilistic and indeterministic nature, effectively made sense of a huge range of peculiarities at the subatomic level, from the way of behaving of electrons in molecules to the operations of quantum fields and particles. Be that as it may, this achievement included some major disadvantages - the hypothesis seemed contradictory with the standards of General Relativity, especially in the portrayal of gravity at vast scopes.

The Conflict of Titans: Relativity versus Quantum Mechanics

The principal conflict between Broad Relativity and quantum mechanics turns out to be most clear while endeavoring to depict the way of behaving of items in outrageous circumstances, like the peculiarity at the core of a dark opening or the minutes soon after the Enormous detonation. General Relativity portrays the gravitational connections in these unique situations, while quantum mechanics oversees the way of behaving of particles. However, these two hypotheses give off an impression of being intrinsically contradictory in such situations.

At the core of the contrariness lies the issue of scale. General Relativity's portrayal of gravity depends on the persistent and smooth shape of spacetime. Conversely, quantum mechanics presents a degree of granularity, proposing that spacetime and matter could have discrete properties at the quantum level. These abberations among consistent and discrete depictions of the universe make irregularities that are especially articulated in circumstances of outrageous gravity, where the old style laws of gravity separate, and quantum impacts become critical.

The mission for a hypothesis that brings together Broad Relativity and quantum mechanics has been a significant undertaking in hypothetical material science. Such a hypothesis, frequently alluded to as a hypothesis of quantum gravity, would give a system that overcomes any issues between the perceptible and tiny domains, permitting us to comprehend the universe from its biggest scales down to its littlest constituents.

A few methodologies have been investigated in the mission for a hypothesis of quantum gravity, each offering an exceptional viewpoint on the best way to accommodate the different universes of relativity and quantum mechanics. These methodologies incorporate string hypothesis, circle quantum gravity, and causal dynamical triangulation, among others.

String Hypothesis: A Vibrational Orchestra of the Universe

String hypothesis is one of the most notable possibility for a hypothesis of quantum gravity. It withdraws from the conventional idea of particles as point-like elements and on second thought places that the central structure blocks of the universe are little vibrating strings. These strings vibrate at various frequencies, bringing about different molecule types in the universe.

String hypothesis offers the possibility to accommodate the contradictions between Broad Relativity and quantum mechanics. At the quantum level, it acquaints a granularity comparable with quantum mechanics, and it integrates the standards of quantum field hypothesis. Be that as it may, at plainly visible scales, string hypothesis is equipped for depicting the consistent and smooth spacetime of General Relativity.

String hypothesis' capability to overcome any barrier between the naturally visible and infinitesimal universes is charming, however it accompanies critical difficulties. One of the significant difficulties is the presence of various string speculations, which shift in the quantity of aspects they expect for consistency. This variety brings up issues about the uniqueness and prescient force of string hypothesis. Besides, string hypothesis presently can't seem to make explicit, tentatively testable expectations, making it hard to confirm its legitimacy.

Circle Quantum Gravity: A Discrete Dance of Quanta

Circle quantum gravity, one more competitor for a hypothesis of quantum gravity, leaves from the ceaseless spacetime of General Relativity and embraces the idea of a discrete, granular spacetime. It proposes that space itself is quantized, made out of inseparable units or "quanta" of space. These quanta structure an organization of interconnected circles, leading to the design of spacetime.

One of the charming parts of circle quantum gravity is capacity to determine a portion of the singularities emerge in Everyday Relativity, like the peculiarity at the focal point of a dark opening. This goal comes from the discrete idea of spacetime, which forestalls the arrangement of vastly thick areas.

Circle quantum gravity offers another viewpoint on the unification of General Relativity and quantum mechanics, integrating quantum standards at both perceptible and tiny scopes. Nonetheless, such as string hypothesis, circle quantum gravity faces difficulties, including its capacity to make tentatively testable forecasts and its similarity with the current group of actual information.

Causal Dynamical Triangulation: Building Spacetime from Triangles

Causal Dynamical Triangulation (CDT) is a one of a kind way to deal with quantum gravity that considers spacetime as an organization of interconnected triangles. These triangles address discrete structure blocks of spacetime that meet up to frame the texture of the universe. CDT means to portray the way of behaving of issue and gravity inside this located spacetime.

One of the eminent highlights of CDT is its capacity to resolve the subject of the bolt of time. By forcing causal circumstances on the triangulation of spacetime, CDT normally integrates the unevenness of time, offering a potential answer for the issue of time's unidirectional stream. This angle separates CDT from different ways to deal with quantum gravity.

While CDT offers a clever viewpoint on the unification of General Relativity and quantum mechanics, it likewise faces difficulties, including the requirement for a steady detailing that can give tentatively testable expectations.

Rising Hypotheses of Quantum Gravity: Another Worldview

Emanant hypotheses of quantum gravity offer an alternate way to deal with overcoming any issues among relativity and quantum mechanics. These speculations suggest that the major laws of physical science, including gravity, may rise up out of additional basic standards at a more profound level. This change in perspective recommends that existence.

5.3 The search for a unified theory of physics

The journey for a brought together hypothesis of material science, frequently alluded to as the "Hypothesis of Everything," has been a focal objective of hypothetical physical science for a really long time. It addresses the undertaking to accommodate and bind together the essential powers and rules that oversee the universe, giving a sound structure that makes sense of all peculiarities, from the way of behaving of subatomic particles to the elements of the universe. In this investigation, we dig into the verifiable setting, the difficulties, and the continuous endeavors in the quest for a brought together hypothesis of physical science.

The Requirement for Solidarity: The Basic Powers of the Universe

The universe is administered by four essential powers: gravity, electromagnetism, the solid atomic power, and the feeble atomic power. Every one of these powers assumes an unmistakable part in forming the actual world, from the gravitational fascination between enormous items to the electromagnetic cooperations that tight spot iotas and particles. The solid atomic power ties protons and neutrons in nuclear cores, while the frail atomic power administers particular kinds of radioactive rots.

Among these powers, gravity and electromagnetism have been portrayed by deeply grounded speculations. General Relativity, created by Albert Einstein in 1915, makes sense of the way of behaving of gravity by demonstrating it as the bend of spacetime.

Electromagnetism is portrayed by Maxwell's situations, formed in the nineteenth hundred years, which offer a complete record of electric and attractive cooperations.

The solid and frail atomic powers, nonetheless, have been trying to bind together with the other two powers. Quantum Chromodynamics (QCD) is the hypothesis that portrays areas of strength for the power, overseeing the cooperations among quarks and gluons. The Electroweak Hypothesis, proposed by Sheldon Glashow, Abdus Salam, and Steven Weinberg, effectively brought together the electromagnetic and frail atomic powers, however these powers stay particular from gravity.

Hypothetical physicists have long tried to foster a brought together hypothesis that integrates every one of the four central powers into a solitary, durable structure. Such a hypothesis wouldn't just give a more profound comprehension of the actual regulations overseeing the universe yet in addition offer the potential for leap forwards in regions like cosmology, molecule physical science, and our cognizance of the early universe.

Authentic Setting: The Way to Unification

The journey for a brought together hypothesis of physical science is established throughout the entire existence of science, where each significant improvement has crawled us nearer to understanding the major rules that oversee the universe. The excursion toward unification has been portrayed by a progression of noteworthy hypotheses and revelations.

One of the earliest cases of unification was the amalgamation of power and attraction by James Agent Maxwell in the nineteenth hundred years. Maxwell's conditions effectively bound together these two peculiarities into a solitary electromagnetic hypothesis, showing that electric and attractive fields were interlaced and engendered as electromagnetic waves, which incorporate light.

Einstein's Exceptional Hypothesis of Relativity, distributed in 1905, laid the basis for the unification of existence, presenting the idea of spacetime as a bound together element. It reclassified the old style ideas of synchronization, time expansion, and the invariance of the speed of light. The hypothesis was a forerunner to the more complete structure of General Relativity, which integrated the power of gravity into the texture of spacetime.

Quantum mechanics, created in the mid twentieth hundred years, reformed how we might interpret the tiny world. It presented the idea of wave-molecule duality, which obscured the limits among particles and waves. Quantum mechanics likewise brought about the Vulnerability Guideline, which forced inborn cutoff points on our capacity to quantify specific properties of particles at the same time. While quantum mechanics effectively depicted the way of behaving of subatomic particles and their cooperations, it presented difficulties in accommodating with General Relativity.

One of the earliest endeavors to bring together powers came from Einstein himself, who looked for a brought together field hypothesis that would include both gravity and electromagnetism. His endeavors, like the Kaluza-Klein hypothesis, meant to insert electromagnetism inside a bound together mathematical system. In any case, these undertakings didn't yield a total hypothesis.

The following significant achievement in the quest for unification accompanied the improvement of the Standard Model of molecule physical science. The Standard Model effectively brought together the electromagnetic and powerless atomic powers inside the system of the Electroweak Hypothesis. This unification prompted the expectation of the W and Z bosons, which were hence found, affirming the hypothesis.

The unification of the solid atomic power with the electroweak force stayed a test, in the long run prompting the definition of the hypothesis of Quantum Chromodynamics (QCD). QCD portrayed areas of strength for the, zeroing in on the cooperations among quarks and gluons, the particles that make up protons, neutrons, and different hadrons.

While these accomplishments stamped huge strides toward unification, the quest for an exhaustive hypothesis that consolidates every one of the four key powers and makes sense of the texture of the universe at all scales stayed progressing.

String Hypothesis: A Promising Structure

One of the most conspicuous and promising possibility for a brought together hypothesis of material science is string hypothesis. String hypothesis withdraws from the old style origination of particles as point-like substances and places that the major structure blocks of the universe are minuscule, vibrating strings. These strings waver at different frequencies, leading to various molecule types in the universe.

String hypothesis offers a convincing structure for unification since it envelops both General Relativity and quantum mechanics. At the quantum level, strings acquaint a granularity comparable with quantum mechanics, and they can be depicted utilizing the standards of quantum field hypothesis. Nonetheless, at plainly visible scales, string hypothesis is fit for portraying the consistent and smooth spacetime of General Relativity.

String hypothesis' ability to overcome any barrier between the naturally visible and minuscule universes has been a main thrust behind its investigation. Inside string hypothesis, there are different forms, including Type I, Type IIA, Type IIB, and heterotic strings. The presence of these various string speculations prompted the improvement of M-hypothesis, which proposes that these different string hypotheses are interconnected and address various features of a more extensive system.

M-hypothesis, as a hypothetical system, offers a way toward unification by proposing that the different string speculations are various indications of a solitary overall hypothesis. This idea gives a method for accommodating the essential powers and rules that oversee the universe.

One of the charming parts of string hypothesis is its capability to depict gravity, one of the essential powers, as a developing property. In this view, gravity rises up out of the aggregate way of behaving of endless strings, each adhering to the laws of string hypothesis. This new point of view on gravity presents a special way to deal with unification and lines up with the idea of gravity as the bend of spacetime in Everyday Relativity.

In any case, in spite of its true capacity, string hypothesis accompanies critical difficulties and questions. One of the essential difficulties is the requirement for exploratory approval. The outrageous scales related with string hypothesis make it trying to lead direct analyses that could affirm or discredit its forecasts. This absence of observational proof has prompted banters inside established researchers about the situation with string hypothesis as a veritable hypothesis of quantum gravity.

Moreover, the huge number of variants of string hypothesis and the intricacy of the arithmetic included have brought up issues about the uniqueness and prescient force of the hypothesis. While string hypothesis has taken huge steps in making sense of specific peculiarities and giving bits of knowledge into the idea of spacetime, it presently can't seem to create tentatively testable forecasts that could conclusively approve or refute the hypothesis.

Notwithstanding these difficulties, string hypothesis stays a conspicuous and convincing competitor in the quest for a bound together hypothesis of physical science. Its capability to overcome any issues between Broad Relativity and quantum mechanics has accumulated critical consideration and continuous examination endeavors.

Circle Quantum Gravity: A Discrete Point of view

Circle quantum gravity is one more competitor in the journey for a brought together hypothesis of material science. This approach withdraws from the consistent spacetime of General Relativity and recommends that space itself is quantized, made out of inseparable units or "quanta" of space. These quanta structure an organization of interconnected circles, leading to the construction of spacetime.

One of the captivating parts of circle quantum gravity is its capacity to resolve the issue of singularities, like the peculiarity at the focal point of a dark opening in Everyday Relativity. The discrete idea of spacetime in circle quantum gravity forestalls the development of limitlessly thick areas, possibly settling these singularities.

Circle quantum gravity offers another point of view on the unification of General Relativity and quantum mechanics by integrating quantum standards at both plainly visible and minuscule scopes. While the hypothesis is still being developed and faces difficulties of its own, it presents a special way to deal with overcoming any barrier between the two principal speculations.

Developing Hypotheses: A Change in outlook

Developing hypotheses of quantum gravity propose an alternate way to deal with unification by recommending that space, time, and even gravity may be emanant properties of a more principal substrate. These hypotheses challenge our old style comprehension of the idea of the universe, recommending that the crucial laws of physical science could emerge from more profound.

Chapter 6

Quantum Mechanics and the Universe

Quantum Mechanics, an essential hypothesis in the domain of material science, has reshaped how we might interpret the universe as well as brought up significant issues about the idea of reality itself. From the subatomic world to the huge universe, the standards of quantum mechanics assume a critical part in our perception of the universe.

At its center, quantum mechanics manages the way of behaving of particles and waves at the littlest scales, administering the way of behaving of iotas, particles, and the central constituents of issue and energy. It arose in the mid twentieth hundred years as a reaction to the limits of traditional physical science, which experienced issues making sense of the way of behaving of particles for these little scopes.

One of the central standards of quantum mechanics is the wave-molecule duality. It declares that particles like electrons and photons can display both molecule like and wave-like qualities relying upon how they are noticed or estimated. This duality challenges our old style instinct, as particles can act as discrete elements with explicit positions, yet additionally as waves with related probabilities.

One more key idea in quantum mechanics is superposition. This standard expresses that quantum frameworks can exist in different states all the while, just imploding into a positive state upon estimation. This property has prompted the improvement of quantum registering, which can possibly change data handling by tackling the force of superposition to perform complex computations at fantastic velocities.

Entrapment is one more cryptic part of quantum mechanics. It recommends that particles can become related so that the estimation of one molecule quickly influences the condition of another, no matter what the distance isolating them. This peculiarity, broadly portrayed by Einstein as "creepy activity a ways off," challenges our traditional comprehension of causality and proposes that data can travel quicker than the speed of light.

The quantum world is represented by probabilistic regulations, as depicted by the Schrödinger condition, which frames how wave capabilities advance in time. Wave capabilities address the likelihood conveyances of a molecule's situation or properties, and they can be utilized to make expectations about the way of behaving of quantum frameworks. Notwithstanding, these expectations are intrinsically probabilistic, presenting a component of vulnerability that is essential to quantum mechanics.

The Vulnerability Rule, figured out by Werner Heisenberg, features as far as possible to our capacity to quantify specific sets of properties, for example, a molecule's situation and energy all the while. This rule suggests that there is a principal breaking point to the accuracy with which we can know these properties, building up the probabilistic idea of quantum mechanics.

The significant ramifications of quantum mechanics stretch out a long ways past the tiny universe of particles and waves. They straightforwardly affect how we might interpret the universe in general. For instance, quantum mechanics assumes a critical part in our perception of the way of behaving of stars, universes, and the whole universe.

In the early snapshots of the universe, during the time of enormous expansion, quantum variances at the littlest scales are accepted to have impacted the huge scope design of the universe. These minuscule variances developed and brought about the appropriation of systems and grandiose microwave foundation radiation we notice today. Quantum mechanics, in this sense, lays the foundation for how we might interpret the astronomical web and the huge scope design of the universe.

Besides, the way of behaving of issue and radiation in outrageous circumstances, like the extraordinary gravitational fields close to dark openings, is represented by the standards of quantum mechanics. The investigation of dark openings has prompted the advancement of quantum field hypothesis in bended spacetime, which is fundamental for how we might interpret how particles and fields cooperate within the sight serious areas of strength for of.

The idea of Peddling radiation, proposed by Stephen Selling, shows the exchange between quantum mechanics and gravity. It proposes that dark openings can discharge particles and radiation because of quantum impacts close to their occasion skylines. This peculiarity challenges our old style comprehension of dark openings as absolutely retaining objects, revealing insight into the multifaceted connection between quantum mechanics and gravity.

Besides, quantum mechanics assumes a focal part in the field of quantum cosmology, which tries to give a quantum depiction of the whole universe. This aggressive undertaking plans to consolidate the standards of quantum mechanics with the hypothesis of general relativity, which depicts gravity on astronomical scales.

One of the difficulties in quantum cosmology is the plan of a quantum hypothesis of gravity, as the ongoing structures of general relativity and quantum mechanics have not yet been effectively brought together. Notwithstanding, the quest for a quantum hypothesis of gravity is fundamental for figuring out the early snapshots of the universe, including the Huge explosion itself. It is accepted that a quantum hypothesis of gravity will give experiences into the idea of singularities and the central condition of the universe before the development of reality.

The quantum idea of the universe additionally becomes possibly the most important factor when we think about the advancement of stars and the course of atomic combination. Stars like the Sun, for instance, depend on quantum burrowing to defeat the electrostatic aversion between emphatically charged nuclear cores and empower the combination of hydrogen into helium. This quantum peculiarity is the motor that powers stars and supports life on The planet.

Quantum mechanics even effects how we might interpret dull matter, a strange and undetectable substance that makes up a huge part of the universe's mass. While dull matter doesn't associate with light or conventional matter similarly, it actually maintains the standards of quantum mechanics, affecting the huge scope design of the universe.

The field of quantum cosmology likewise dives into the idea of the multiverse, which is a speculative thought recommending the presence of numerous universes past our own. With regards to quantum mechanics, the multiverse speculation emerges from the possibility of quantum superposition. It sets that each time a quantum occasion has numerous potential results, the universe branches into various equal real factors, each comparing to an alternate result. While the multiverse speculation is right now dubious and exceptionally speculative, it brings up significant issues about the idea of the universe and our place inside it.

Quantum mechanics challenges our old style instincts, and its suggestions stretch out even to the philosophical and magical domains. The idea of wave capability breakdown, for example, brings up issues about the idea of the real world and the job of awareness in estimation. The well known Schrödinger's feline psychological study, which includes a feline in a superposition of alive and dead states, features the quirks of quantum estimation and its association with the plainly visible world.

The understanding of quantum mechanics stays a subject of discussion among physicists and rationalists. There are different translations, like the Copenhagen understanding, the many-universes understanding, and the pilot-wave hypothesis.

Each offers an alternate point of view on the idea of quantum reality, and the decision of translation has significant ramifications for how we might interpret the universe.

The Copenhagen translation, proposed by Niels Bohr and Werner Heisenberg, recommends that the demonstration of estimation or perception makes the wave capability breakdown, yielding an unequivocal result. This understanding puts a focal job on the spectator and brings up issues about the idea of reality free of perception.

The many-universes translation, then again, suggests that all potential results of a quantum occasion happen in discrete, non-imparting parts of the universe. This understanding suggests that each conceivable history and future exists as discrete equal real factors, offering an answer for the issue of wave capability breakdown.

Nonetheless, it presents the test of making sense of how these branches stay inconspicuous and hush.

The pilot-wave hypothesis, created by Louis de Broglie and David Bohm, presents the idea of stowed away factors that decide the clear condition of a quantum framework consistently. In this view, particles have distinct positions and speeds, yet their way of behaving is impacted by a pilot wave that directs their movement. This understanding plans to save determinism and authenticity in quantum mechanics however isn't without its own difficulties.

The philosophical ramifications of quantum mechanics likewise reach out to the idea of causality, determinism, and freedom of thought. The probabilistic idea of quantum occasions difficulties the old style thought of a precision universe, where each cause makes a determinate difference. All things being equal, quantum mechanics presents a component of arbitrariness and vulnerability, bringing up issues about the constraints of expectation and the job of chance in the universe.

Moreover, the idea of ensnarement challenges how we might interpret circumstances and logical results. On the off chance that particles can become caught and their properties are connected paying little heed to separate, it recommends a take-off from the old style idea of neighborhood causality, where an occasion at one area can influence close by occasions. The peculiarity of ensnarement infers a more interconnected and non-nearby perspective on the universe.

The ramifications of quantum mechanics likewise stretch out to the way of thinking of psyche. The job of cognizance in the estimation cycle stays a subject of philosophical discussion. A few translations of quantum mechanics, like the Copenhagen understanding, recommend that cognizance assumes a focal part in wave capability breakdown. This brings up issues about the idea of cognizance and its relationship to the actual world.

The investigation of quantum mechanics has prompted innovative headways that have changed our lives. Quantum mechanics supports the standards of semiconductors and strong state physical science, which are fundamental for the advancement of electronic gadgets and PCs.

Quantum mechanics likewise assumes a significant part in the field of quantum optics, prompting the improvement of lasers and current broadcast communications.

Quantum registering is a state of the art field that saddles the standards of quantum mechanics to perform computations that would be unimaginable for traditional PCs. Quantum PCs can possibly reform cryptography, improvement issues, and the recreation of complicated quantum frameworks, with applications in fields like medication disclosure and material science.

Quantum instant transportation, one more quantum peculiarity, has been shown in the research center and holds guarantee for secure correspondence and data move. While it doesn't include the prompt transportation of issue, quantum instant transportation takes into consideration the exchange of quantum states starting with one area then onto the next, offering another worldview for quantum correspondence.

Quantum mechanics has likewise led to the field of quantum cryptography, which use the standards of quantum superposition and ensnarement to get correspondence channels. Quantum key conveyance frameworks offer a degree of safety that is hypothetically strong, as any endeavor to snoop on the correspondence would disturb the sensitive quantum states being communicated.

Besides, the investigation of quantum mechanics has prompted the advancement of new materials with noteworthy properties. For example, superconductors, which can convey electric flow with zero obstruction, have tracked down applications in clinical imaging, transportation, and energy transmission. Quantum dabs, nanoscale semiconductor particles, have been used in quantum specks shows and photovoltaic cells, empowering more energy-effective lighting and sun based energy age.

Quantum mechanics has likewise assumed an essential part in the improvement of quantum sensors and imaging advances. Quantum-improved sensors can accomplish phenomenal degrees of accuracy in fields, for example, gravitational wave location, attractive reverberation imaging, and nuclear clocks. These progressions have broad ramifications for basic exploration and functional applications.

In the field of quantum optics, tests have affirmed the unusual and entrancing forecasts of quantum mechanics. Tests including trapped photons have shown the infringement of Ringer's disparities, supporting the view that quantum ensnarement is a genuine and non-traditional peculiarity. The improvement of quantum optics has additionally prompted the acknowledgment of quantum instant transportation and quantum cryptography in the lab.

The investigation of quantum mechanics has likewise had suggestions for how we might interpret time and the idea of the bolt of time.

The idea of time balance in quantum mechanics recommends that the basic laws of physical science are symmetric concerning the heading of time. Be that as it may, the perceptible world seems to show a one-way bolt of time, portrayed by the increment of entropy and the irreversibility of specific cycles.

The rise of thermodynamics from the hidden standards of quantum mechanics has revealed insight into the beginning of the bolt of time. The second law of thermodynamics, which expresses that entropy will in general increment after some time, is personally associated with the probabilistic way of behaving of particles at the quantum level. While quantum mechanics itself is time-reversible, the measurable way of behaving of huge gatherings of particles leads to the bolt of time we see in our daily existences.

The investigation of quantum mechanics has additionally prompted significant inquiries concerning the idea of reality. With regards to quantum cosmology, the job of reality as major substances is tested. The plan of a quantum hypothesis of gravity intends to give a depiction of the universe that doesn't depend on previous spacetime however rather rises out of additional key standards.

String hypothesis, a hypothetical structure that plans to bind together broad relativity and quantum mechanics, presents the idea of additional aspects past the recognizable three spatial aspects and once aspect. String hypothesis proposes that the key structure blocks of the universe are not particles but rather small strings, vibrating at various frequencies. This hypothesis gives a possible road to accommodating the unique structures of quantum mechanics and general relativity.

The ramifications of quantum mechanics for the idea of the universe are both significant and expansive. They challenge our traditional instincts, bring up issues about the idea of the real world and causality, and motivate continuous philosophical and logical discussions. Quantum mechanics has not just upset how we might interpret the subatomic world yet has likewise made a permanent imprint on our understanding of the universe at large.

All in all, quantum mechanics remains as quite possibly of the most significant and secretive hypothesis throughout the entire existence of science. Its standards, which oversee the way of behaving of particles and waves at the littlest scales, have reshaped how we might interpret the universe from the subatomic domain to the actual universe. The wave-molecule duality, superposition, and ensnarement are among the central ideas that challenge our traditional instinct and keep on filling logical investigation and philosophical request.

The ramifications of quantum mechanics reach out past the lab and the domain of physical science. They address the idea of causality, determinism, and the job of cognizance in estimation.

The different understandings of quantum mechanics, from the Copenhagen translation to the many-universes speculation, offer alternate points of view on the idea of quantum reality, bringing up significant issues about the idea of the universe and our place inside it.

Besides, quantum mechanics has prompted innovative progressions that have changed our lives, from electronic gadgets and PCs to quantum figuring and quantum cryptography. It has opened up new boondocks in material science and imaging advancements, empowering the improvement of superconductors, quantum dabs, and quantum-improved sensors.

The investigation of quantum mechanics has likewise affected how we might interpret time and the bolt of time. The association between quantum conduct and the increment of entropy in the plainly visible world gives experiences into the beginning of the bolt of time we see in our daily existences.

At last, the investigation of quantum mechanics has significant ramifications for how we might interpret the universe, from the way of behaving of particles and waves at the quantum level to the development of reality with regards to quantum cosmology. It moves our old style instincts and keeps on rousing logical request and philosophical reflection, leaving us with a profound feeling of marvel and interest in the idea of the real world and the universe.

6.1 Quantum cosmology and the early moments of the universe

Quantum cosmology dives into the core of the universe's starting point and early minutes, planning to blend the standards of quantum mechanics with the hypothesis of general relativity. This aggressive field of study looks to respond to significant inquiries concerning the actual idea of the universe, its beginning, and the key condition of the universe before the development of reality.

The universe, as we notice it today, is immense, dynamic, and loaded up with systems, stars, and inestimable designs. Be that as it may, our vast excursion starts at a lot more modest and denser scale, during a period known as the Huge explosion. This occasion, which happened roughly 13.8 quite a while back, denoted the introduction of the universe as far as we might be concerned.

Quantum cosmology places that the universe, at its earliest minutes, existed in a condition of outrageous thickness and energy, where the impacts of both quantum mechanics and general relativity were similarly huge. This exceptional inestimable climate opposes our traditional instincts, as it is a domain where the standards of the tiny (quantum mechanics) and the extremely enormous (general relativity) meet.

The major system of quantum cosmology is established in the hypothesis of general relativity, created by Albert Einstein in the mid twentieth hundred years.

General relativity portrays the power of gravity as the shape of spacetime by gigantic items, and it has effectively made sense of the elements of the universe for vast scopes, from the way of behaving of worlds to the design of the universe overall.

Notwithstanding, when we rewind the vast clock to the absolute starting point, to the thick and hot states of the Enormous detonation, general relativity experiences its cutoff points. At this outrageous energy scale, the impacts of quantum mechanics become huge. The converging of these two central hypotheses is the critical test of quantum cosmology.

One of the focal inquiries quantum cosmology addresses is the means by which to portray the universe's state right now of the Huge explosion. In the system of general relativity, the universe seems to rise up out of a peculiarity, a place of boundless thickness and shape. In any case, this peculiarity addresses a breakdown of our actual hypotheses, as related with vast qualities show the deficiency of our ongoing comprehension.

Quantum cosmology plans to determine this issue by giving a quantum portrayal of the universe's underlying state. This includes regarding the whole universe as a quantum framework and applying the standards of quantum mechanics to comprehend how it developed from a quantum state into the universe we notice today.

The presentation of quantum mechanics into cosmology requires a change in context. Instead of imagining the universe as a deterministic framework represented exclusively by the laws of traditional material science, quantum cosmology embraces a probabilistic and indeterministic view. In this quantum system, the condition of the universe at the Huge explosion is portrayed by a wave capability, which embodies the probabilities of various conceivable vast setups.

The improvement of quantum cosmology has prompted the definition of different models and speculations that endeavor to depict the early snapshots of the universe. One of the most noticeable methodologies is known as the Wheeler-DeWitt condition, named after physicists John Archibald Wheeler and Bryce DeWitt. This condition fills in as a foundation in the field of quantum cosmology and endeavors to exemplify the quantum conduct of the whole universe.

The Wheeler-DeWitt condition, basically, addresses the quantum simple of Hamilton's rule from traditional mechanics. It portrays the development of the universe's wave capability concerning time, consolidating both the standards of general relativity and quantum mechanics. This condition gives experiences into the quantum conduct of the universe, its development, and the elements of existence.

The Wheeler-DeWitt condition and other quantum cosmological models have offered new viewpoints on the early snapshots of the universe. They recommend that the universe might not have had a clear start at the Huge explosion peculiarity however rather rose up out of a quantum state with a scope of conceivable beginning circumstances.

This view difficulties the thought of a "creation occasion" and makes the way for the likelihood that the universe has consistently existed in some structure.

Quantum cosmology likewise investigates the idea of enormous expansion, which sets that the universe went through a fast and outstanding development at the times following the Huge explosion. Inflationary hypothesis, proposed by physicist Alan Guth, gives a possible clarification to the consistency and enormous scope design of the universe. It proposes that quantum vacillations at the littlest scales during expansion brought about the thickness varieties that cultivated the development of systems and astronomical designs.

With regards to quantum cosmology, grandiose expansion is seen as a quantum cycle driven by the elements of a scalar field, known as the inflaton. This scalar field, impacted by the standards of quantum mechanics, prompts the remarkable extension of the universe, streamlining abnormalities and making way for the development of worlds and inestimable designs.

The quantum idea of expansion brings up issues about a definitive reason for the universe's development. In old style cosmology, the development of the universe is frequently credited to the underlying peculiarity at the Enormous detonation. Nonetheless, in a quantum cosmological structure, the universe's extension is a consequence of quantum vacillations and the probabilistic way of behaving of the inflaton field.

The investigation of quantum cosmology additionally brings up significant issues about the idea of time. In traditional material science, time is treated as an outright and free boundary that streams consistently, giving a decent structure to the portrayal of occasions. Be that as it may, the quantum portrayal of the universe challenges this old style thought of time.

In quantum cosmology, time turns into a new idea, complicatedly connected to the elements of the actual universe. The Wheeler-DeWitt condition and other quantum cosmological models recommend that time, similar to space, may have started from the quantum condition of the universe. This point of view lines up with the more extensive idea of emanant spacetime, which sets that the actual texture of spacetime emerges from more principal quantum levels of opportunity.

The rise of time with regards to quantum cosmology has prompted fascinating thoughts regarding the idea of the universe's "bolt of time." The bolt of time alludes to the one-way course wherein situation transpire, portrayed by the increment of entropy and the irreversibility of specific cycles.

The second law of thermodynamics, a crucial rule of traditional physical science, expresses that entropy, a proportion of turmoil, will in general increment over the long haul. This regulation is personally associated with the probabilistic way of behaving of particles at the quantum level.

While quantum mechanics itself is time-reversible, the factual way of behaving of enormous gatherings of particles leads to the bolt of time we see in our daily existences.

The ramifications of quantum cosmology stretch out to the idea of singularities, areas of spacetime where the bend turns out to be limitlessly enormous, as on account of the Huge explosion peculiarity. In old style general relativity, singularities are tricky, as they address a breakdown of the hypothesis and demonstrate the requirement for a more major portrayal.

Quantum cosmology offers a possible goal to the issue of singularities by presenting a quantum depiction of the universe's initial minutes. Instead of a mark of boundless thickness, the Enormous detonation peculiarity might be supplanted by a quantum state with a limited, but very high, energy thickness. This viewpoint lines up with the possibility that the actual idea of a peculiarity is tested by the standards of quantum mechanics.

The idea of a "skip" is one more interesting part of quantum cosmology. In specific quantum cosmological models, the universe might not have a particular start at the Huge explosion, yet all things being equal, it goes through a pattern of development and constriction. This cyclic universe situation, proposed by scholars like Paul Steinhardt and Neil Turok, recommends that the universe could be in a timeless pattern of astronomical birth and demise.

In this cyclic model, the universe's development arrives at a most extreme degree, after which it contracts, just to extend again in a ceaseless cycle. The quantum idea of the universe assumes a urgent part in this situation, as it presents the chance of keeping away from singularities and giving a nonstop depiction of the universe.

The investigation of quantum cosmology has likewise led to speculative thoughts regarding the multiverse. The multiverse speculation proposes the presence of various universes past our own, each with its own arrangement of actual regulations and astronomical circumstances. With regards to quantum cosmology, the multiverse emerges from the possibility of quantum superposition.

As indicated by this speculation, each time a quantum occasion has numerous potential results, the universe branches into various equal real factors, each comparing to an alternate result. While the multiverse speculation is right now doubtful and exceptionally speculative, it brings up significant issues about the idea of the universe and our place inside it.

The ramifications of quantum cosmology stretch out to the more extensive journey for a hypothesis of everything, a bound together system that would accommodate the principal powers of the universe, specifically gravity, electromagnetism, and the solid and frail atomic powers. The hypothesis of everything is a well established objective in hypothetical material science, and quantum cosmology assumes a urgent part in this undertaking.

String hypothesis, a hypothetical structure that expects to bring together the powers of nature and depict the central structure blocks of the universe, has been impacted by the experiences of quantum.

6.2 The role of quantum reality in the evolution of the cosmos

The investigation of quantum the truth is in a general sense entwined with how we might interpret the universe and its development. Quantum mechanics, which oversees the way of behaving of particles and waves at the littlest scales, assumes a vital part in forming the universe from its initial minutes to its present status. This complex exchange between the quantum world and inestimable development gives a significant point of view on the idea of reality itself.

At its center, quantum mechanics challenges our traditional instincts by presenting the idea of wave-molecule duality. This guideline recommends that particles like electrons and photons can show both molecule like and wave-like qualities relying upon how they are noticed or estimated. At the end of the day, particles are not confined to acting as discrete elements with explicit positions yet can likewise show wave-like properties with related probabilities.

Wave-molecule duality has critical ramifications for how we might interpret the universe. It infers that at the quantum level, the actual idea of issue and energy is portrayed by a principal vulnerability. This vulnerability is embodied in the wave works that depict the probabilities of a molecule's situation, force, and different properties.

Quantum superposition is one more focal idea in quantum mechanics. It expresses that quantum frameworks can exist in different states at the same time until an estimation is made, so, all in all the framework falls into a distinct state. Superposition lies at the core of quantum registering, a field that saddles the force of numerous quantum states to perform complex estimations dramatically quicker than traditional PCs.

Quantum trap is one more charming peculiarity of quantum reality. It proposes that particles can become corresponded so that the estimation of one molecule promptly influences the condition of another, regardless of whether they are isolated by tremendous distances. This "creepy activity a ways off," as Albert Einstein broadly portrayed it, challenges our traditional comprehension of causality and proposes that data can be traded quicker than the speed of light.

The quantum world is represented by probabilistic regulations, as depicted by the Schrödinger condition, which frames how wave capabilities advance in time. These wave capabilities address the likelihood disseminations of a molecule's situation or properties and can be utilized to make forecasts about the way of behaving of quantum frameworks. In any case, these expectations are intrinsically probabilistic, presenting a component of vulnerability that is essential to quantum mechanics.

The Vulnerability Rule, planned by Werner Heisenberg, is another key idea. It features as far as possible to our capacity to gauge specific sets of properties, for example, a molecule's situation and energy all the while. This guideline suggests that there is a central cutoff to the accuracy with which we can know these properties, building up the probabilistic idea of quantum mechanics.

The standards of quantum reality become much more charming when we think about their job in the development of the universe. The universe, as far as we can tell, started with the Huge explosion around 13.8 quite a while back. At that early second, the whole universe existed in a condition of outrageous thickness and energy. In this early stage climate, quantum mechanics and general relativity, the hypothesis of gravity, were both similarly critical.

The converging of quantum mechanics and general relativity at the introduction of the universe presents significant inquiries concerning the idea of astronomical advancement. While general relativity depicts the elements of the universe for huge scopes, including the way of behaving of worlds and the development of the universe, it experiences its cutoff points when we dive into the thick and vigorous states of the Enormous detonation.

Quantum mechanics, then again, becomes basic at the quantum scale, where particles and fields interface as indicated by probabilistic regulations and display wave-molecule duality. The test of quantum cosmology is to give a bound together depiction of the universe that consolidates both the standards of quantum mechanics and general relativity.

One of the focal inquiries that quantum cosmology addresses is the manner by which to depict the universe's state right now of the Enormous detonation. In traditional cosmology, the universe rises out of a peculiarity, a place of endless thickness and bend. In any case, this peculiarity addresses a breakdown of our actual hypotheses, as related with vast qualities show the deficiency of our ongoing comprehension.

Quantum cosmology plans to determine this issue by giving a quantum depiction of the universe's underlying state. This includes regarding the whole universe as a quantum framework and applying the standards of quantum mechanics to comprehend how it developed from a quantum state into the universe we notice today.

In the system of quantum cosmology, the universe is portrayed by a wave capability that exemplifies the probabilities of various conceivable grandiose setups. This wave capability, like those of particles in the quantum domain, addresses the probabilistic appropriation of the universe's properties, like its size, energy, and arch.

Quite possibly of the most unmistakable methodology in quantum cosmology is the Wheeler-DeWitt condition, named after physicists John Archibald Wheeler and Bryce DeWitt.

This condition fills in as a foundation in the field of quantum cosmology and endeavors to embody the quantum conduct of the whole universe.

The Wheeler-DeWitt condition is the quantum simple of Hamilton's standard, a major idea in traditional mechanics. It portrays the development of the universe's wave capability concerning time, consolidating both the standards of general relativity and quantum mechanics. This condition gives experiences into the quantum conduct of the universe, its development, and the elements of reality.

The Wheeler-DeWitt condition and other quantum cosmological models offer fascinating experiences into the early snapshots of the universe. They propose that the universe might not have had an unequivocal start at the Huge explosion peculiarity, as traditional cosmology recommends. All things considered, the universe might have risen up out of a quantum state with a scope of conceivable starting circumstances.

This view difficulties the thought of a "creation occasion" and makes the way for the likelihood that the universe has consistently existed in some structure, with its initial minutes portrayed by a quantum superposition of various potential states. In this unique situation, the universe's introduction to the world is definitely not a solitary occasion yet rather a progress from a quantum state to a particular traditional reality.

Quantum cosmology likewise investigates the idea of enormous expansion, a hypothesis that recommends the universe went through a fast and remarkable extension at the times following the Huge explosion. Inflationary hypothesis, proposed by physicist Alan Guth, gives a possible clarification to the consistency and huge scope construction of the universe. It recommends that quantum changes at the littlest scales during expansion brought about the thickness varieties that cultivated the development of universes and enormous designs.

With regards to quantum cosmology, grandiose expansion is seen as a quantum interaction driven by the elements of a scalar field, known as the inflaton. This scalar field, impacted by the standards of quantum mechanics, prompts the dramatic development of the universe, streamlining abnormalities and making way for the arrangement of worlds and enormous designs.

The quantum idea of expansion brings up issues about a definitive reason for the universe's extension. In traditional cosmology, the development of the universe is frequently credited to the underlying peculiarity at the Enormous detonation.

Notwithstanding, in a quantum cosmological system, the universe's development is a consequence of quantum changes and the probabilistic way of behaving of the inflaton field.

Quantum cosmology offers a remarkable viewpoint on the rise of time. In traditional material science, time is treated as a flat out and free boundary that streams consistently, giving a proper system to the depiction of occasions. Notwithstanding, the quantum portrayal of the universe challenges this old style idea of time.

In quantum cosmology, time turns into a new idea, unpredictably connected to the elements of the actual universe. The Wheeler-DeWitt condition and other quantum cosmological models recommend that time, similar to space, may have started from the quantum condition of the universe. This viewpoint lines up with the more extensive idea of new spacetime, which places that the actual texture of spacetime emerges from more crucial quantum levels of opportunity.

The rise of time with regards to quantum cosmology has prompted charming thoughts regarding the idea of the universe's "bolt of time." The bolt of time alludes to the one-way bearing where situation transpire, described by the increment of entropy and the irreversibility of specific cycles.

The second law of thermodynamics, a key standard of old style material science, expresses that entropy, a proportion of confusion, will in general increment over the long run. This regulation is personally associated with the probabilistic way of behaving of particles at the quantum level. While quantum mechanics itself is time-reversible, the measurable way of behaving of huge gatherings of particles leads to the bolt of time we see in our regular daily existences.

The ramifications of quantum cosmology reach out to the idea of singularities, areas of spacetime where the shape turns out to be limitlessly huge, as on account of the Enormous detonation peculiarity. In traditional general relativity, singularities are tricky, as they address a breakdown of the hypothesis and show the requirement for a more major depiction.

6.3 Dark matter, dark energy, and their connection to quantum physics

The universe, insofar as we can tell, is an intricate and strange spot. While the noticeable matter, including worlds, stars, and planets, comprises just a little part of the universe, a critical piece stays concealed in the shadows, as dim matter and dim energy. The presence and properties of these mysterious parts are integral to our understanding of the universe, and their association with quantum material science adds one more layer of interest to the vast account.

Dull matter is a conjectured type of issue that doesn't produce, ingest, or connect with electromagnetic radiation, like light. Its presence was first proposed by Swiss space expert Fritz Zwicky during the 1930s when he saw that the apparent matter in world bunches couldn't represent the noticed gravitational powers keeping these groups intact. The expression "dull matter" itself insinuates its tricky nature, as it stays imperceptible and imperceptible through conventional means.

The proof for dull matter's presence has developed further throughout the long term, as space experts and astrophysicists significantly affect the movements of worlds and the enormous scope construction of the universe. Dim matter is remembered to make up generally 27% of the universe's absolute mass and energy, overshadowing the 5% addressed by noticeable matter. The excess 68% of the universe's energy financial plan is credited to dull energy, a significantly more puzzling and strange part.

Dull energy, in contrast to dim matter, isn't a type of issue by any stretch of the imagination yet rather a reasonable placeholder for an obscure power that has all the earmarks of being speeding up the development of the universe. The presence of dim energy was first proposed in the last part of the 1990s, when perceptions of far off supernovae uncovered that the universe's extension isn't dialing back, as one could expect because of gravitational fascination, however is, truth be told, accelerating.

The idea of dim energy stays quite possibly of the main riddle in present day cosmology. It isn't related with any known molecule or field in the standard model of molecule physical science, and its source and properties stay speculative. Dim energy is described by its negative strain, which prompts a loathsome gravitational impact, making the universe's development speed up.

The association between dull matter, dim energy, and quantum physical science lies in the key standards of quantum mechanics, which support how we might interpret matter and energy at the littlest scales. To fathom this association, it is fundamental to dive into the quantum world and investigate how it connects with the strange parts that shape the universe.

In quantum mechanics, the way of behaving of particles is depicted by wave capabilities, numerical portrayals that typify the probabilities of different properties, like position, force, and energy. Wave capabilities give a probabilistic structure to figuring out the way of behaving of particles, as they permit us to foresee the probability of different results when estimations are made.

Wave-molecule duality, one of the focal standards of quantum mechanics, proposes that particles, like electrons and photons, can show both molecule like and wave-like qualities relying upon how they are noticed or estimated. This standard infers that the properties of particles are innately unsure until an estimation is made, so, all in all the wave capability falls into a positive state.

The probabilistic idea of quantum mechanics challenges our old style instincts, as it presents a component of inborn vulnerability into the principal conduct of issue and energy. In the quantum world, particles don't have obvious positions and momenta until they are noticed, and their properties are portrayed concerning probabilities.

The idea of superposition is one more major part of quantum mechanics. It expresses that quantum frameworks can exist in various states at the same time until an estimation is made. This guideline, exemplified by Schrödinger's popular psychological study including a feline in a superposition of life and demise, features the fascinating and frequently outlandish nature of quantum reality.

Superposition has suggestions for how we might interpret matter and energy at the grandiose scale, especially with regards to dull matter. While dim matter remaining parts slippery and imperceptible, its properties are construed through gravitational impacts. It is remembered to comprise of a yet-unseen class of particles that don't collaborate with electromagnetic powers, making them challenging to straightforwardly recognize.

In the quantum domain, particles like those proposed for dim matter can exist in superpositions of various states. This implies that dull matter particles, in principle, could all the while possess various positions and speeds, prompting a probabilistic dispersion of their thickness all through the universe. The idea of superposition gives a quantum structure to grasping the dissemination and conduct of dull matter in the universe.

Quantum mechanics likewise assumes a part in the development and development of grandiose designs, for example, systems and world groups. The probabilistic way of behaving of issue in the early universe, impacted by quantum vacillations, led to the thickness varieties that cultivated the arrangement of grandiose designs. These vacillations, engraved in the enormous microwave foundation radiation, act as a fossilized record of the quantum processes that formed the universe.

The association between dull matter and quantum physical science turns out to be much more clear when we think about the idea of dim matter particles. While the particular character of dull matter particles stays obscure, different competitors have been proposed, including Feebly Collaborating Monstrous Particles (Weaklings) and Axions. These particles, assuming they exist, would be administered by the standards of quantum mechanics.

Weaklings, for example, are anticipated to be gigantic particles that collaborate just feebly with other matter, subsequently their name. They are supposed to be their own antiparticles and would keep the guidelines of quantum field hypothesis, a structure that consolidates quantum mechanics and extraordinary relativity. The properties and conduct of Weaklings, as quantum particles, would be portrayed by wave works and dependent upon quantum vulnerability.

Quantum field hypothesis, which underlies the way of behaving of rudimentary particles, is fundamental for figuring out the quantum properties of dim matter competitors. The collaborations and properties of these particles, for example, their dispersing and demolition processes, are portrayed with regards to quantum field hypothesis. This system permits physicists to make expectations about how dim matter might connect with apparent matter and how it could be distinguished in tests.

Dim matter locators, for example, those situated in profound underground research facilities, try to catch the associations of dull matter particles with conventional matter. These examinations depend on the probabilistic idea of quantum mechanics, as they intend to distinguish uncommon occasions that happen when dim matter particles, on the off chance that they exist, cooperate with the finders' nuclear cores.

The association between quantum physical science and dim matter additionally stretches out to the quest for new physical science past the standard model of molecule physical science. The standard model, which portrays the central particles and powers of the universe, is exceptionally fruitful yet has constraints. It doesn't represent the presence of dull matter, and it likewise neglects to consolidate the power of gravity.

Quantum gravity, a hypothetical structure that looks to join the standards of quantum mechanics and general relativity, is fundamental for grasping the way of behaving of issue and energy at the inestimable scale. While the standard model doesn't give a total portrayal of quantum gravity, the journey for a brought together hypothesis that envelops both quantum mechanics and gravity stays a focal test in hypothetical physical science.

String hypothesis, a hypothetical structure that means to bind together the powers of nature and portray the essential structure blocks of the universe, is one of the main contender for a hypothesis of quantum gravity. In string hypothesis, particles are supplanted by little strings that vibrate at various frequencies, bringing about the different particles and powers known to man.

String hypothesis presents the idea of additional aspects past the recognizable three spatial aspects and once aspect. These additional aspects, compactified at tiny scopes, assume a part in the quantum conduct of issue and energy. String hypothesis, as a quantum structure, can possibly give bits of knowledge into the idea of dim matter and dim energy and their association with the key standards of the universe.

The idea of dull energy, which stays quite possibly of the main secret in cosmology, likewise has associations with quantum material science. Dim energy is portrayed by its negative strain, which prompts a ghastly gravitational impact, making the universe's development speed up. While the idea of dim energy isn't surely known, it is believed to be related with a field or energy that pervades space itself.

In quantum field hypothesis, the vacuum state isn't unfilled however is loaded up with quantum vacillations and virtual particles that constantly arise and obliterate. These vacuum changes are a key part of quantum mechanics and lead to the idea of vacuum energy. With regards to dull energy, a few speculations propose that the vacuum energy related with quantum variances could add to the frightful power liable for the universe's sped up extension.

The grandiose association between dull matter, dim energy, and quantum material science is a demonstration of the significant transaction between the essential rules that oversee the universe. While dull matter remaining parts an imperceptible and tricky .

Chapter 7

Quantum Technologies and Their Impact

The universe, as far as we can tell, is a mind boggling and strange spot. While the noticeable matter, including worlds, stars, and planets, is just a little part of the universe, a huge piece stays concealed in the shadows, as dim matter and dull energy. The presence and properties of these confounding parts are vital to our perception of the universe, and their association with quantum material science adds one more layer of interest to the inestimable account.

Dull matter is a speculated type of issue that doesn't emanate, retain, or communicate with electromagnetic radiation, like light. Its presence was first proposed by Swiss cosmologist Fritz Zwicky during the 1930s when he saw that the apparent matter in world bunches couldn't represent the noticed gravitational powers keeping these groups intact. The expression "dim matter" itself insinuates its subtle nature, as it stays imperceptible and imperceptible through conventional means.

The proof for dim matter's presence has developed further throughout the long term, as space experts and astrophysicists affect the movements of cosmic systems and the huge scope design of the universe. Dull matter is remembered to make up generally 27% of the universe's complete mass and energy, overshadowing the 5% addressed by apparent matter. The excess 68% of the universe's energy financial plan is credited to dull energy, a considerably more cryptic and baffling part.

Dim energy, in contrast to dull matter, isn't a type of issue by any means yet rather a calculated placeholder for an obscure power that gives off an impression of being speeding up the extension of the universe. The presence of dim energy was first proposed in the last part of the 1990s, when perceptions of far off supernovae uncovered that the universe's development isn't dialing back, as one could expect because of gravitational fascination, however is, as a matter of fact, accelerating.

The idea of dim energy stays quite possibly of the main riddle in current cosmology. It isn't related with any known molecule or field in the standard model of molecule physical science, and its source and properties stay speculative. Dull energy is portrayed by its negative tension, which prompts an unpleasant gravitational impact, making the universe's development speed up.

The association between dull matter, dim energy, and quantum material science lies in the central standards of quantum mechanics, which support how we might interpret matter and energy at the littlest scales. To appreciate this association, it is fundamental to dive into the quantum world and investigate how it connects with the puzzling parts that shape the universe.

In quantum mechanics, the way of behaving of particles is portrayed by wave capabilities, numerical portrayals that typify the probabilities of different properties, like position, force, and energy. Wave capabilities give a probabilistic system to grasping the way of behaving of particles, as they permit us to foresee the probability of different results when estimations are made.

Wave-molecule duality, one of the focal standards of quantum mechanics, proposes that particles, like electrons and photons, can show both molecule like and wave-like attributes relying upon how they are noticed or estimated. This guideline suggests that the properties of particles are innately questionable until an estimation is made, so, all in all the wave capability implodes into a positive state.

The probabilistic idea of quantum mechanics challenges our old style instincts, as it presents a component of innate vulnerability into the essential way of behaving of issue and energy. In the quantum world, particles don't have distinct positions and momenta until they are noticed, and their properties are depicted regarding probabilities.

The idea of superposition is one more principal part of quantum mechanics. It expresses that quantum frameworks can exist in various states all the while until an estimation is made. This guideline, exemplified by Schrödinger's renowned psychological study including a feline in a superposition of life and demise, features the fascinating and frequently strange nature of quantum reality.

Superposition has suggestions for how we might interpret matter and energy at the astronomical scale, especially with regards to dull matter. While dim matter remaining parts subtle and imperceptible, its properties are deduced through gravitational impacts. It is remembered to comprise of a yet-unseen class of particles that don't co-operate with electromagnetic powers, making them challenging to straightforwardly recognize.

In the quantum domain, particles like those proposed for dull matter can exist in superpositions of numerous states. This implies that dull matter particles, in principle, could all the while possess various positions and speeds, prompting a probabilistic dissemination of their thickness all through the universe. The idea of superposition gives a quantum structure to grasping the dispersion and conduct of dull matter in the universe.

Quantum mechanics likewise assumes a part in the development and advancement of enormous designs, for example, universes and cosmic system bunches. The probabilistic way of behaving of issue in the early universe, affected by quantum vacillations, led to the thickness varieties that cultivated the development of enormous designs. These changes, engraved in the enormous microwave foundation radiation, act as a fossilized record of the quantum processes that formed the universe.

The association between dim matter and quantum material science turns out to be much more clear when we think about the idea of dull matter particles. While the particular character of dull matter particles stays obscure, different applicants have been proposed, including Feebly Interfacing Huge Particles (Weaklings) and Axions. These particles, in the event that they exist, would be administered by the standards of quantum mechanics.

Weaklings, for example, are anticipated to be huge particles that cooperate just pitifully with other matter, consequently their name. They are supposed to be their own antiparticles and would keep the guidelines of quantum field hypothesis, a system that joins quantum mechanics and exceptional relativity. The properties and conduct of Weaklings, as quantum particles, would be portrayed by wave works and dependent upon quantum vulnerability.

Quantum field hypothesis, which underlies the way of behaving of rudimentary particles, is fundamental for grasping the quantum properties of dim matter up-and-comers. The cooperations and properties of these particles, for example, their dissipating and destruction processes, are depicted with regards to quantum field hypothesis. This system permits physicists to make forecasts about how dim matter might communicate with noticeable matter and how it could be recognized in tests.

Dim matter locators, for example, those situated in profound underground labs, try to catch the connections of dull matter particles with customary matter. These analyses depend on the probabilistic idea of quantum mechanics, as they plan to recognize uncommon occasions that happen when dim matter particles, assuming they exist, cooperate with the indicators' nuclear cores.

The association between quantum physical science and dull matter likewise stretches out to the quest for new physical science past the standard model of molecule physical science. The standard model, which portrays the major particles and powers of the universe, is profoundly effective yet has constraints. It doesn't represent the presence of dull matter, and it likewise neglects to consolidate the power of gravity.

Quantum gravity, a hypothetical system that looks to join the standards of quantum mechanics and general relativity, is fundamental for grasping the way of behaving of issue and energy at the inestimable scale. While the standard model doesn't give a total depiction of quantum gravity, the journey for a bound together hypothesis that envelops both quantum mechanics and gravity stays a focal test in hypothetical physical science.

String hypothesis, a hypothetical system that plans to bind together the powers of nature and depict the major structure blocks of the universe, is one of the main contender for a hypothesis of quantum gravity. In string hypothesis, particles are supplanted by minuscule strings that vibrate at various frequencies, leading to the different particles and powers known to man.

String hypothesis presents the idea of additional aspects past the recognizable three spatial aspects and once aspect. These additional aspects, compactified at tiny scopes, assume a part in the quantum conduct of issue and energy. String hypothesis, as a quantum system, can possibly give experiences into the idea of dim matter and dim energy and their association with the crucial standards of the universe.

The idea of dim energy, which stays quite possibly of the main secret in cosmology, additionally has associations with quantum material science. Dim energy is portrayed by its negative tension, which prompts a horrible gravitational impact, making the universe's development speed up. While the idea of dim energy isn't surely known, it is believed to be related with a field or energy that penetrates space itself.

In quantum field hypothesis, the vacuum state isn't unfilled yet is loaded up with quantum variances and virtual particles that consistently arise and obliterate. These vacuum vacillations are a basic part of quantum mechanics and lead to the idea of vacuum energy. With regards to dim energy, a few speculations propose that the vacuum energy related with quantum vacillations could add to the ghastly power liable for the universe's sped up extension.

The inestimable association between dull matter, dim energy, and quantum physical science is a demonstration of the significant transaction between the basic rules that oversee the universe. While dim matter remaining parts an imperceptible and tricky .

7.1 Quantum technologies that are reshaping our world

Quantum innovations address an extraordinary outskirts in science and designing, promising to reform different parts of our lives. At their center, these advancements influence the standards of quantum mechanics, which oversee the way of behaving of issue and energy at the littlest scales. As quantum innovations keep on propelling, their effect on fields like processing, correspondence, detecting, and cryptography is turning out to be progressively significant. In this thorough investigation, we'll dive into the groundworks of quantum mechanics, the standards hidden quantum advancements, and their extensive ramifications for what's in store.

Quantum mechanics, frequently alluded to as quantum physical science or quantum hypothesis, is the part of physical science that depicts the way of behaving of issue and energy at the nuclear and subatomic scale. Created in the mid twentieth 100 years, quantum mechanics addresses a change in perspective from traditional physical science, which portrays the way of behaving of ordinary items. At the quantum level, particles, for example, electrons and photons show wave-molecule duality, and that implies they can at the same time display both molecule like and wave-like qualities. This duality is epitomized in the idea of wave capabilities, numerical portrayals that depict the probabilities of different properties, like position, force, and energy.

Superposition, one more essential idea of quantum mechanics, expresses that quantum frameworks can exist in various states all the while until an estimation is made, so, all in all the framework falls into a clear state. This property leads to the charming and frequently irrational nature of quantum reality. Superposition permits quantum bits, or qubits, to exist in a mix of 0 and 1 states, as opposed to solely as either, as in old style bits. This quality is vital to quantum figuring and assumes a urgent part in quantum innovations.

Trap is a peculiarity in quantum mechanics that connects the properties of particles so that the estimation of one molecule momentarily influences the condition of another, regardless of whether they are isolated by huge distances. This "creepy activity a ways off," as Albert Einstein broadly portrayed it, challenges old style thoughts of causality and recommends the presence of mind boggling connections in the quantum domain.

Quantum innovations outfit these and other quantum peculiarities to empower remarkable advances in a scope of fields, starting with quantum figuring. Quantum processing is maybe the most notable utilization of quantum advances and can possibly change computational power. Dissimilar to traditional pieces, which can exist as 0 or 1, qubits can exist in superpositions of 0 and 1 states, permitting quantum PCs to at the same time investigate numerous potential arrangements. This parallelism guarantees dramatic speedup for specific computational undertakings, like calculating huge numbers and taking care of perplexing enhancement issues.

One of the most celebrated quantum calculations is Shor's calculation, which can effectively factor huge numbers — an undertaking considered very trying for old style PCs. Calculating huge numbers is the premise of numerous encryption strategies, and Shor's calculation represents a likely danger to old style encryption methods. Quantum PCs likewise hold guarantee in fields like medication revelation, material science, and enhancement issues with boundless commonsense applications.

While quantum PCs are still in the beginning phases of advancement, organizations like IBM, Google, and new companies like Rigetti are gaining critical headway. Quantum incomparability, the place where a quantum PC outflanks traditional PCs on unambiguous errands, has been accomplished for specific issues. As quantum equipment keeps on improving, the effect on fields like cryptography, man-made reasoning, and logical examination is supposed to be significant.

Quantum correspondence is one more groundbreaking part of quantum advances. Quantum key dispersion (QKD) empowers secure correspondence by utilizing the standards of quantum mechanics to make a cryptographic key that is hypothetically solid. QKD use the peculiarity of snare to guarantee that any listening in on the quantum correspondence channel would be quickly identified. This degree of safety has huge ramifications for safeguarding delicate data in regions like government, money, and medical services.

Quantum detecting is an arising area of quantum innovations that use the accuracy of quantum mechanics to make exceptionally delicate sensors. For example, quantum sensors can identify minute changes in attractive fields, gravitational powers, and electromagnetic radiation. These sensors have applications in a large number of fields, including topography, natural checking, and clinical imaging.

Quantum metrology, a subfield of quantum detecting, is especially encouraging for accomplishing accuracy estimations. The upgraded responsiveness of quantum sensors empowers more exact estimations of actual amounts, which can fundamentally affect logical exploration and industry. For instance, in gravitational wave identification, quantum sensors can possibly work on the responsiveness and precision of estimations, eventually prompting a more profound comprehension of grandiose peculiarities.

Quantum imaging is a developing area of quantum innovations that offers further developed imaging and detecting capacities. Quantum imaging use quantum trap and the wave-molecule duality of quantum particles to make pictures with upgraded highlights. This innovation can be applied in fields like microscopy, remote detecting, and clinical imaging, offering more significant subtlety and responsiveness in catching pictures.

Quantum cryptography is a subset of quantum correspondence that spotlights on secure encryption strategies in view of the standards of quantum mechanics. Quantum key dissemination (QKD), referenced prior, is a focal part of quantum cryptography. It gives an on a very basic level secure method for trading encryption keys, making listening in essentially unimaginable. Quantum cryptography can possibly improve the security of advanced correspondence, especially for exceptionally touchy data.

Quantum recreations are a strong utilization of quantum innovations for understanding and reenacting complex quantum frameworks. Quantum test systems can impersonate the way of behaving of quantum materials and frameworks, which are trying to concentrate on utilizing traditional PCs. This capacity is especially significant in fields like dense matter physical science, science, and materials science, where quantum communications assume a urgent part.

The field of quantum innovations keeps on advancing quickly, and its effect on different businesses and logical fields is now significant. Quantum innovations hold the possibility to address difficulties that were already unconquerable for old style draws near. As quantum equipment and calculations advance, we can hope to see quantum advances assume an undeniably persuasive part in our lives.

In any case, it's fundamental to recognize that quantum advances likewise present one of a kind difficulties. Quantum PCs, while strong, are presently restricted in their capacities and face impediments connected with mistake rectification and versatility. Guaranteeing the security of quantum correspondence and cryptography is a continuous exertion, as expected weaknesses and assaults on quantum frameworks should be entirely perceived and relieved.

Besides, quantum innovations are capital-concentrated and require particular skill, making them open to a predetermined number of associations and scientists. The turn of events and mix of quantum advancements into different businesses will require proceeded with speculations and joint efforts across the scholarly world, industry, and government.

Taking everything into account, quantum advancements are ready to reshape the scene of calculation, correspondence, detecting, and security. Quantum registering holds the possibility to change computational power, while quantum correspondence offers rugged encryption strategies. Quantum detecting and imaging give upgraded accuracy and responsiveness, and quantum cryptography guarantees secure advanced correspondence. Quantum reproductions empower the investigation of perplexing quantum frameworks, helping fields like materials science and science.

While quantum advancements are still in their beginning phases, their effect is as of now being felt across various areas. As progressions in quantum equipment and calculations proceed, the extraordinary capability of these advancements will turn out to be considerably more evident. The exchange between the essential standards of quantum mechanics and their commonsense applications in quantum advances high-lights the profundity of their effect on our lives, offering new capacities and answers for longstanding difficulties.

7.2 Quantum computing and its potential applications

Quantum processing remains at the cutting edge of a mechanical transformation, ready to reshape the scene of calculation and take care of perplexing issues that have long escaped traditional PCs. Utilizing the standards of quantum mechanics, quantum PCs can possibly beat their old style partners in different spaces, from cryptography and streamlining to sedate disclosure and materials science. This investigation dives into the essentials of quantum processing, the present status of the field, and the potential applications that guarantee to introduce another time of registering.

Quantum processing is essentially not the same as old style registering. While old style PCs use bits, which can address either a 0 or a 1, quantum PCs use qubits, which can exist in a superposition of 0 and 1 states. This implies that quantum PCs can all the while investigate numerous potential answers for an issue, taking advantage of the force of parallelism to play out specific calculations dramatically quicker.

The idea of superposition is one of the center standards of quantum registering. A qubit can exist in a superposition of both 0 and 1, which permits quantum calculations to handle data in a profoundly equal and probabilistic way. This quality is fundamental to quantum figuring's capacity to handle complex issues, as it empowers the investigation of different arrangements on the double.

Entrapment, one more crucial quantum peculiarity, is likewise saddled in quantum registering. When qubits are entrapped, the estimation of one qubit momentarily impacts the condition of the other, regardless of whether they are isolated by immense distances. Ensnarement takes into consideration the formation of mind boggling quantum expresses that empower quantum PCs to perform multifaceted activities.

Quantum registering isn't just a quicker variant of old style processing; it addresses a change in perspective in calculation. Quantum calculations, like Shor's calculation and Grover's calculation, have shown the possibility to take care of issues that are thought of as recalcitrant for old style PCs. Shor's calculation, for instance, productively factors enormous numbers, representing a danger to traditional encryption techniques.

One of the most notable possible utilizations of quantum processing is in the area of cryptography. Current encryption techniques, which depend on the trouble of considering enormous numbers or tackling complex numerical issues, could be defenseless against quantum assaults. Quantum PCs, with their capacity to effectively factor enormous numbers, might actually break generally utilized encryption plans, like RSA.

Post-quantum cryptography is a field that looks to foster encryption strategies that are secure against quantum assaults. It includes the plan of cryptographic calculations that would stay strong even within the sight of strong quantum PCs. This work is basic to guarantee the security of computerized correspondence and data assurance in a quantum-controlled world.

Quantum registering's effect on cryptography isn't restricted to breaking encryption. It additionally offers the possibility to improve security through the advancement of quantum-safe encryption techniques. Quantum key circulation (QKD), a part of quantum cryptography, gives an on a very basic level secure method for trading encryption keys, making snoopping essentially unthinkable. The security given by QKD has huge ramifications for safeguarding touchy data in regions like government, money, and medical services.

Quantum registering can possibly upset drug disclosure and advancement. One of the most computationally serious parts of medication disclosure is reenacting and breaking down the way of behaving of atoms and compound responses. Quantum PCs can essentially speed up these recreations, empowering scientists to investigate a more extensive scope of sub-atomic collaborations and foster new medications all the more rapidly and productively.

For instance, quantum PCs can recreate complex sub-atomic designs and quantum frameworks, which is fundamental for understanding medication collaborations at the quantum level. This ability can prompt the disclosure of novel mixtures and more successful medications for different ailments, from malignant growth to irresistible sicknesses.

Quantum registering additionally holds guarantee in the field of materials science. Materials researchers look to find and plan new materials with explicit properties, for example, superconductors for effective energy transmission or high level materials for aviation applications. Quantum PCs can mimic the way of behaving of materials at the quantum level, permitting specialists to distinguish promising applicants and improve material properties.

The capacity to precisely show quantum frameworks and connections gives an upper hand in materials science. Quantum PCs offer a stage for recreating the electronic design of materials and foreseeing their properties with phenomenal accuracy. This empowers the revelation of cutting edge materials that can change ventures and advances.

Enhancement issues, which include tracking down the best arrangement among countless conceivable outcomes, are unavoidable in different spaces, from operations and money to assembling and transportation. Old style PCs battle with complex enhancement issues, as they require broad computational assets to investigate every single expected arrangement.

Quantum registering succeeds in improvement undertakings because of its characteristic parallelism and the capacity to all the while investigate various arrangements. Quantum calculations like Grover's calculation and quantum strengthening give proficient ways of looking through arrangement spaces and track down ideal or close ideal arrangements. This capacity has commonsense applications in production network improvement, portfolio the board, and coordinated factors, to give some examples.

Quantum AI is an arising field that consolidates quantum registering and computerized reasoning (simulated intelligence). It intends to use quantum calculations and quantum information portrayal to improve AI undertakings, like example acknowledgment, information investigation, and advancement. Quantum AI can possibly speed up man-made intelligence applications, making them all the more remarkable and proficient.

Quantum calculations like the quantum support vector machine (QSVM) and quantum brain networks offer benefits in taking care of mind boggling AI issues. Quantum information can be encoded in quantum states, which can address a lot of data productively. This ability is especially valuable for handling enormous information and working on the exactness of man-made intelligence models.

Quantum reproductions are a fundamental utilization of quantum registering in different logical disciplines. Quantum PCs can reenact the way of behaving of quantum frameworks and materials with unmatched exactness. This ability is important for research in dense matter material science, quantum science, and high-energy physical science.

Quantum test systems give experiences into the way of behaving of quantum particles and the properties of intricate materials. They permit scientists to investigate quantum peculiarities, for example, high-temperature superconductivity and quantum stage changes, which have wide ramifications in science and innovation.

Quantum information examination and quantum-improved AI offer novel ways to deal with taking care of immense datasets and figuring out complex data. Quantum PCs can proficiently process and investigate enormous datasets by exploiting quantum parallelism and superposition. This ability has applications in fields like information examination, design acknowledgment, and direction.

Quantum AI, specifically, is ready to disturb the field of information examination. Quantum calculations can reveal stowed away examples, separate experiences from loud information, and streamline dynamic cycles. This is particularly pertinent in regions where information examination assumes a urgent part, like money, medical care, and logical exploration.

Quantum correspondence, as a use of quantum processing, intends to make secure correspondence channels by utilizing the standards of quantum mechanics. Quantum key conveyance (QKD), which gives an in a general sense secure method for trading encryption keys, is at the front of quantum correspondence research. QKD offers tough encryption strategies, as any snoopping on the quantum correspondence channel would be quickly identified.

Secure correspondence is fundamental in different spaces, including government, money, medical services, and safeguard. Quantum correspondence vows to upgrade the security of computerized correspondence and shield delicate data from digital dangers. Its effect stretches out to get casting a ballot frameworks, monetary exchanges, and information security.

Quantum advancements, including quantum processing, are additionally opening new outskirts in the investigation of quantum peculiarities and principal physical science. Tests including quantum snare and superposition are propelling comprehension we might interpret the quantum domain. Besides, quantum advancements are adding to the improvement of novel quantum sensors and imaging gadgets that offer phenomenal accuracy in logical estimations.

For instance, quantum sensors can distinguish and gauge actual amounts, like attractive fields and gravitational powers, with unrivaled awareness. These sensors have applications in fields like topography, natural observing, and clinical imaging. Quantum imaging, which takes advantage of quantum properties to catch improved pictures, is gaining ground in microscopy, remote detecting, and logical examination.

Quantum metrology, a subfield of quantum detecting, intends to accomplish accuracy estimations past the capacities of old style instruments. The upgraded responsiveness of quantum sensors considers the precise estimation of actual boundaries, which has suggestions in logical examination, industry, and medical care.

7.3 Quantum communication and cryptography

Quantum correspondence and quantum cryptography address state of the art innovations that influence the standards of quantum mechanics to empower secure and solid correspondence. These fields are at the very front of guaranteeing the secrecy and trustworthiness of advanced data in an undeniably interconnected world. In this investigation, we dig into the underpinnings of quantum correspondence and cryptography, the improvement of quantum key circulation, and the expected applications and difficulties of these quantum advancements.

Traditional correspondence depends on the trading of data utilizing old style bits, which can address either a 0 or a 1. In old style cryptography, calculations are utilized to encode and decipher messages, and the security of the encryption depends on the trouble of tackling explicit numerical issues, like figuring huge numbers or tracking down discrete logarithms. Traditional encryption strategies, while compelling, are possibly helpless against assaults by strong PCs, including quantum PCs.

Quantum correspondence and cryptography, then again, depend on the standards of quantum mechanics, which administer the way of behaving of particles at the nuclear and subatomic scale. Quantum correspondence use quantum properties, like superposition and ensnarement, to make in a general sense secure correspondence channels. The most notable use of quantum correspondence is quantum key appropriation (QKD), which empowers the trading of encryption keys with provable security.

One of the crucial standards basic quantum correspondence is the idea of superposition. In quantum mechanics, particles like electrons and photons can exist in superpositions of numerous states at the same time. For instance, a quantum bit, or qubit, can exist in a superposition of 0 and 1 states, permitting it to address the two qualities simultaneously. This property empowers quantum correspondence to investigate numerous potential outcomes at the same time, which has significant ramifications for encryption and security.

Ensnarement is one more fundamental rule in quantum correspondence. At the point when two qubits become entrapped, the estimation of one qubit quickly decides the condition of the other, regardless of whether they are isolated by huge distances. This peculiarity, as broadly portrayed by Albert Einstein as "creepy activity a good ways off," has critical ramifications for secure correspondence.

The idea of quantum trap is key to quantum key dissemination (QKD). In a QKD convention, two gatherings, customarily alluded to as Alice and Sway, make a caught sets of qubits. By estimating their particular qubits, they can produce a common mystery key. If a busybody, known as Eve, endeavors to catch the qubits during transmission, the trap property guarantees that any snoopping would unavoidably change the quantum condition of the qubits. This change would be quickly recognized, making quantum correspondence uncommonly secure.

Quantum key appropriation (QKD) is the leader use of quantum correspondence, offering provably secure encryption keys for secure computerized correspondence. QKD uses the standards of quantum mechanics to make encryption keys that are hypothetically solid, even by a strong quantum PC. The security of QKD depends on the difficulty of cloning erratic quantum expresses, a peculiarity known as the "no-cloning hypothesis."

The most notable QKD convention is the BB84 convention, created by Charles Bennett and Gilles Brassard in 1984. In the BB84 convention, Alice sends a surge of qubits to Sway. These qubits can be in one of four states, which compare to two symmetrical bases. Alice openly reports the bases utilized for encoding, and Sway records the bases he utilizes for estimation. This data trade permits Alice and Bounce to figure out which qubits they can use to make a common mystery key. The no-cloning hypothesis guarantees that any snooping endeavor by Eve would present blunders that are perceptible by Alice and Weave.

The security of QKD conventions is ensured by the major standards of quantum mechanics, especially the no-cloning hypothesis and the property of snare. The discovery of snooping depends on the way that any estimation performed on a quantum state unavoidably changes its quantum properties, making it unthinkable for a busybody to catch the qubits without identification.

QKD has pragmatic applications in getting touchy data across different areas. In the field of government and guard, QKD is utilized to safeguard ordered correspondences and public safety data.

Monetary organizations depend on QKD to get exchanges and shield monetary information. Medical care associations use QKD to safeguard patient records and clinical exploration.

While QKD offers provable security, understanding its limitations is fundamental. Reasonable QKD frameworks are dependent upon different difficulties, including the impacts of channel commotion, signal weakening, and finder defects. Specialists keep on dealing with creating powerful QKD frameworks that can work in genuine circumstances and convey elite execution security.

Quantum correspondence innovations stretch out past QKD to offer secure and effective methods for trading data. Quantum instant transportation, a peculiarity roused by the standards of quantum trap, empowers the exchange of quantum states starting with one area then onto the next. Quantum instant transportation has suggestions for the protected transmission of quantum data, for example, quantum encryption keys.

Quantum repeaters are gadgets intended to broaden the scope of secure quantum correspondence. Over significant distance correspondence channels, quantum signals will more often than not experience the ill effects of sign misfortune and debasement. Quantum repeaters assist with moderating these issues by enhancing and safeguarding quantum data as it traversed huge distances. The advancement of quantum repeaters is a functioning area of exploration in quantum correspondence.

Quantum correspondence can likewise work with secure democratic frameworks. Quantum casting a ballot frameworks use quantum properties to guarantee the respectability and privacy of the democratic interaction. These frameworks can forestall messing with votes and give an elevated degree of confidence in political decision results. Quantum casting a ballot can possibly upgrade the security and straightforwardness of majority rule processes.

Quantum correspondence can have critical ramifications in the field of secure democratic frameworks. Quantum casting a ballot frameworks use quantum properties to ensure the trustworthiness and secrecy of the democratic interaction. These frameworks can forestall vote altering and guarantee a serious level of confidence in political race results. Quantum casting a ballot can possibly upgrade the security and straightforwardness of vote based processes.

Quantum correspondence isn't without its difficulties. One of the main useful issues is the requirement for particular foundation to help quantum correspondence. This foundation incorporates quantum key dispersion (QKD) gadgets, quantum repeaters, and secure quantum channels. Creating and keeping up with this foundation can be costly and actually requesting.

Quantum correspondence frameworks additionally face difficulties connected with signal misfortune and corruption over significant distances. Quantum signs can become weakened and adulterated as they travel through optical filaments or free space. Creating productive quantum repeaters is fundamental to broaden the scope of secure quantum correspondence.

One more test is the possible weakness of quantum correspondence frameworks to assaults that exploit shortcomings in the actual execution. For instance, side-channel assaults, which exploit data spillage from the actual equipment, might possibly think twice about security of quantum correspondence. Analysts and designers are ceaselessly attempting to address these weaknesses and work on the security of quantum correspondence frameworks.

Quantum correspondence frameworks should likewise battle with viable restrictions, for example, the requirement for cooling and seclusion to keep up with the fragile quantum conditions of qubits. These prerequisites can make quantum correspondence frameworks cumbersome and expensive, which can restrict their far and wide sending.

Also, quantum correspondence frameworks should work in conditions with low degrees of commotion and impedance to keep up with the respectability of quantum states. Accomplishing this degree of natural control can be trying in genuine situations.

Quantum cryptography isn't restricted to getting traditional correspondence yet reaches out to quantum-secure correspondence and the improvement of cutting edge cryptographic conventions. Quantum-safe cryptography means to plan encryption strategies that are secure against assaults by strong quantum PCs. As quantum registering keeps on propelling, the requirement for quantum-safe cryptographic arrangements turns out to be progressively critical.

Post-quantum cryptography, otherwise called quantum-safe cryptography, centers around creating encryption strategies that are secure against quantum assaults. These techniques depend on numerical issues that are accepted to be hard in any event, for quantum PCs. Present quantum cryptography points on guarantee the drawn out security of computerized correspondence, especially in situations where delicate data should be safeguarded.

One of the essential ways to deal with post-quantum cryptography is grid based cryptography. Grid put together cryptography depends with respect to the hardness of cross section issues, like the Most brief Vector Issue (SVP) and the Learning With Blunders (LWE) issue. These issues are accepted to be impervious to quantum assaults, making grid based cryptography a promising contender for quantum-safe encryption.

Chapter 8

Interpretations of Quantum Mechanics

Translations of Quantum Mechanics have been a wellspring of significant philosophical and logical discussion starting from the commencement of quantum hypothesis in the mid twentieth hundred years. While quantum mechanics has been phenomenally effective in making sense of the way of behaving of the minuscule world, it brings up issues about the idea of the real world, the job of estimation, and the essential standards of the universe that keep on testing our comprehension. In this conversation, we will investigate a portion of the critical translations of quantum mechanics, each offering an alternate point of view on the most proficient method to decipher the numerical formalism and the peculiarities it portrays.

One of the most generally acknowledged understandings of quantum mechanics is the Copenhagen translation, created by Niels Bohr and Werner Heisenberg during the 1920s. As indicated by this view, the quantum world is on a very basic level probabilistic, and the wave capability, which portrays the condition of a quantum framework, possibly implodes into an unmistakable result when estimated. All in all, the demonstration of estimation assumes a focal part in quantum mechanics, and until an estimation is made, a quantum framework exists in a superposition of potential states.

The Copenhagen translation has been tremendously effective in making sense of trial results and giving a functional structure to quantum physical science. It stresses the significance of the eyewitness and the indeterminacy of quantum frameworks, where certain properties, like an electron's situation and energy, can't be all the while known with inconsistent accuracy — a standard known as the Heisenberg Vulnerability Guideline.

Be that as it may, the Copenhagen understanding has additionally been the subject of extraordinary philosophical discussion. A few pundits contend that it neglects to give an unmistakable ontological structure to figuring out the idea of the quantum world. It brings up issues about what befalls a framework before it is estimated and whether there is a fundamental reality that exists freely of perception.

In light of these worries, elective translations of quantum mechanics have arisen. The Many-Universes Understanding, proposed by Hugh Everett III in 1957, recommends that each conceivable result of a quantum estimation happens, yet every result exists in a different, non-conveying part of the universe. This understanding suggests that the universe continually branches into numerous equal real factors, and all potential results exist together in various branches. It gives a deterministic system where the wave capability never falls, and the spectator simply follows one of numerous potential ways through a steadily growing multiverse.

The Many-Universes Translation has acquired prominence among a smart physicists and offers a method for staying away from the "estimation issue" intrinsic in the Copenhagen understanding. In any case, it raises its own arrangement of philosophical inquiries, for example, the idea of these equal real factors and the issue of how the "parting" of the universe happens.

Another translation, known as the Pilot-Wave Hypothesis or Bohmian mechanics, was created by Louis de Broglie and David Bohm during the 1950s. This understanding proposes that quantum particles are directed by a secret pilot wave that decides their directions. As such, the molecule's situation and energy are not in a general sense questionable still up in the air by the wave capability and the underlying states of the framework. This view gives a deterministic portrayal of quantum mechanics and jelly stowed away factors that decide the results of estimations.

Bohmian mechanics has been the subject of continuous exploration and discussion. While it gives an unmistakable ontological structure, it brings up issues about the idea of the secret factors and how they collaborate with quantum frameworks. Moreover, it presents the idea of non-region, where particles can momentarily impact each other's way of behaving a ways off, testing the standards of unique relativity.

An alternate point of view on quantum mechanics comes from the Conditional Understanding, proposed by John Cramer during the 1980s. This understanding perspectives quantum connections as a two-step process including both high level (future-coordinated) and hindered (past-coordinated) waves. At the point when a producer delivers an impeded wave, it ventures out in reverse so as to meet a high level wave, making a standing wave that addresses an exchange between the producer and safeguard. The Value-based Translation gives a special method for understanding quantum snare and the breakdown of the wave capability.

The Value-based Understanding offers a clever way to deal with quantum peculiarities, underscoring the job of correspondence through cutting edge and hindered waves. Be that as it may, it isn't generally so broadly embraced as different translations and stays a subject of continuous examination and discussion.

Another translation, the Goal Breakdown Models, recommends that the wave capability implodes suddenly because of an actual interaction. In contrast to the Copenhagen understanding, where estimation is vital to wave capability breakdown, these models propose that the breakdown happens autonomously of perception. Different systems have been proposed to make sense of this unconstrained breakdown, fully intent on overcoming any issues between the quantum and traditional universes.

One such model is the Constant Unconstrained Limitation (CSL) hypothesis, which places that particles experience irregular, nonstop falls, prompting a dispersion of their wave capabilities. This model expects to make sense of why naturally visible articles, which are made out of endless quantum particles, display traditional way of behaving and positive properties. Objective Breakdown Models stand out as they give a likely goal to the estimation issue and proposition testable expectations through tests pointed toward recognizing unconstrained breakdowns.

The Social Understanding, proposed via Carlo Rovelli, takes a generally alternate point of view on quantum mechanics by stressing the social idea of actual peculiarities. As indicated by this view, there are no outright properties or conditions of quantum frameworks. All things being equal, the properties of a quantum object are characterized corresponding to another item or eyewitness. This translation challenges the thought of genuine reality and features the significance of the eyewitness' viewpoint in grasping the quantum world.

The Social Translation has created significant interest among savants and physicists, as it offers a new viewpoint on the idea of the real world and the job of perception in quantum mechanics. It lines up with a more extensive philosophical pattern known as relationalism, which challenges the possibility of outright reality.

As of late, Quantum Bayesianism or QBism has acquired consideration as a translation that spotlights on the job of the spectator in quantum mechanics. QBism recommends that the quantum wave capability addresses a specialist's emotional convictions about a quantum framework's results. In this view, quantum mechanics is viewed as a structure for refreshing one's very own levels of conviction in light of proof and perceptions.

QBism puts the spectator at the focal point of quantum hypothesis and rethinks the wave capability as an instrument for Bayesian likelihood thinking. It offers an extreme change in context, stressing that quantum mechanics is a hypothesis about the individual encounters and convictions of eyewitnesses.

The Many-Personalities Translation, proposed by Adrian Kent, is another understanding that spotlights on the spectator's job. In this view, each estimation result makes a new "mind," and the quantum wave capability addresses the conveyance of psyches in various states. It recommends that every onlooker sees a solitary, clear result, however all potential results exist in the "scene of brains."

The Many-Personalities Translation interfaces quantum mechanics to the idea of eyewitnesses and their encounters. While it offers an interesting point of view, it brings up issues about the idea of these "personalities" and how they cooperate with the actual world.

The Quantum Bayesianism and Many-Personalities Understandings highlight the continuous discussion about the idea of quantum estimations and the job of the eyewitness. They challenge the possibility of an objective reality and recommend that quantum mechanics might be a structure for portraying individual encounters and convictions.

An alternate way to deal with understanding quantum mechanics comes from the possibility of Quantum Data Hypothesis. This viewpoint sees quantum states as transporters of data and underlines the job of quantum trap in quantum correspondence and calculation. Quantum Data Hypothesis has prompted the advancement of quantum innovations, for example, quantum cryptography and quantum registering, which outfit the exceptional properties of quantum frameworks for useful applications.

8.1 The various interpretations of quantum mechanics

The Different Understandings of Quantum Mechanics have been a subject of extraordinary discussion and hypothesis starting from the origin of quantum hypothesis in the mid twentieth hundred years. While quantum mechanics has shown to be a strikingly effective structure for understanding the way of behaving of the infinitesimal world, it is joined by an innate philosophical test — the translation of what's going on at the quantum level. This has prompted the improvement of different understandings, each offering an unmistakable point of view on the major idea of quantum reality, the job of estimation, and the way of behaving of quantum frameworks.

One of the most generally perceived understandings of quantum mechanics is the Copenhagen translation, which was figured out by Niels Bohr and Werner Heisenberg during the 1920s. As indicated by this translation, the quantum world is innately probabilistic. It declares that a quantum framework, depicted by a numerical element known as the wave capability, exists in a superposition of potential states until an estimation is made, so, all in all the wave capability implodes to a particular result. Basically, the demonstration of estimation assumes a central part in quantum mechanics, and until noticed, quantum frameworks don't have positive properties.

The Copenhagen translation has been strikingly effective in making sense of exploratory outcomes and giving a useful structure to quantum material science. It puts huge accentuation on the job of the spectator and the innate indeterminacy of quantum frameworks, an idea typified by Heisenberg's Vulnerability Standard. This guideline directs that specific sets of properties, like an electron's situation and force, can't be all the while estimated with erratic accuracy. In any case, while the Copenhagen translation is successful by and by, it has incited significant philosophical discussion. Pundits contend that it neglects to give a reasonable ontological structure to figuring out the idea of the quantum world. It brings up issues about what befalls a quantum framework before it is estimated and whether there is a fundamental reality that exists freely of perception.

Because of these difficulties, a few elective understandings of quantum mechanics have arisen. The Many-Universes Translation, proposed by Hugh Everett III in 1957, adopts a strikingly unique strategy. It proposes that each conceivable result of a quantum estimation really happens, however every result exists in a different, non-imparting part of the universe. In this view, the universe continually parts into various equal real factors, and all possible results exist together in various branches. The Many-Universes Translation offers a deterministic system where the wave capability never falls, and the onlooker follows one of various potential ways inside a consistently extending multiverse.

The Many-Universes Understanding has gathered consideration among certain physicists, as it offers an answer for the "estimation issue" innate in the Copenhagen translation. In any case, it presents its own arrangement of philosophical requests, for example, the idea of these equal real factors and the mechanics of universe parting.

Another fascinating translation is the Pilot-Wave Hypothesis, otherwise called Bohmian mechanics, created by Louis de Broglie and David Bohm during the 1950s. This understanding sets that quantum particles are directed by a disguised pilot wave that decides their directions. Rather than the Copenhagen translation, which depicts specific properties as generally questionable until estimated, Bohmian mechanics presents a deterministic structure where the molecule's situation and force are represented by the wave capability and introductory circumstances.

Bohmian mechanics has its own continuous conversations and discussions. While it gives an unmistakable ontological structure, it brings up issues about the idea of the secret factors and how they communicate with quantum frameworks. Furthermore, it presents the idea of non-territory, where particles can quickly impact each other's way of behaving a ways off, which challenges the standards of extraordinary relativity.

The Conditional Understanding, proposed by John Cramer during the 1980s, offers a one of a kind point of view on quantum communications. It places that quantum exchanges include both high level (future-coordinated) and impeded (past-coordinated) waves. At the point when a producer delivers an impeded wave, it ventures out in reverse so as to meet a high level wave, making a standing wave that addresses an exchange between the producer and safeguard. The Conditional Translation gives an unmistakable method for understanding quantum entrapment and the breakdown of the wave capability.

This understanding highlights the possibility of correspondence through cutting edge and hindered waves. Be that as it may, it isn't so broadly embraced as different understandings and stays a subject of progressing examination and discussion.

Objective Breakdown Models, like Ceaseless Unconstrained Confinement (CSL), recommend that the wave capability suddenly implodes because of an actual interaction. These models propose that the breakdown happens autonomously of perception and give different systems to make sense of this unconstrained breakdown. They expect to overcome any barrier between the quantum and old style universes and proposition testable forecasts through tests intended to distinguish unconstrained breakdowns.

The Social Understanding, presented via Carlo Rovelli, challenges the idea of an objective reality and stresses the social idea of actual peculiarities. As per this view, properties of a quantum object are characterized comparable to another item or onlooker, and there are no outright properties or states. This translation lines up with a more extensive philosophical pattern known as relationalism, which questions the idea of outright reality.

Quantum Bayesianism, or QBism, has acquired unmistakable quality as a translation that puts the onlooker at the focal point of quantum hypothesis. QBism proposes that the quantum wave capability addresses a specialist's emotional convictions about a quantum framework's results. In this view, quantum mechanics is viewed as a system for refreshing one's very own levels of conviction in light of proof and perceptions.

The Many-Personalities Translation, proposed by Adrian Kent, shares an emphasis on the spectator's job. In this view, every estimation result makes a new "mind," and the quantum wave capability addresses the circulation of psyches in various states. The Many-Personalities Translation proposes that every spectator sees a solitary, distinct result, while all potential results exist in the "scene of brains."

Both Quantum Bayesianism and the Many-Personalities Understanding feature the continuous discussion about the idea of quantum estimations and the job of the spectator. They challenge the possibility of an objective reality and propose that quantum mechanics might act as a structure for depicting individual encounters and convictions.

A particular point of view on quantum mechanics emerges from the field of Quantum Data Hypothesis. This perspective considers quantum states as transporters of data and stresses the job of quantum snare in quantum correspondence and calculation. Quantum Data Hypothesis has prompted the improvement of useful quantum advances, for example, quantum cryptography and quantum registering, which exploit the exceptional properties of quantum frameworks for different applications.

Quantum Data Hypothesis has presented a better approach for survey quantum mechanics, stressing the job of data and correspondence in quantum frameworks. It has commonsense applications in fields like secure correspondence and calculation, further obscuring the line between the quantum and old style universes.

The Understandings of Quantum Mechanics address a different scope of viewpoints on the essential inquiries of quantum physical science. They address issues connected with the idea of the real world, the job of the onlooker, and the association between the quantum and old style universes. While every translation has its own benefits and difficulties, the continuous discussion and exploration in this field keep on revealing insight into the secretive and captivating nature of quantum mechanics. As how we might interpret the quantum world extends, these translations will keep on advancing, offering new experiences into the key functions of the universe.

8.2 Comparing the Copenhagen interpretation, many-worlds hypothesis, and more

The field of quantum mechanics is set apart by its profound and generally expected puzzling philosophical ramifications. At its center, quantum mechanics offers a numerical system that depicts the way of behaving of particles at the infinitesimal scale. Be that as it may, the translations of this system have ignited extraordinary discussion and keep on testing how we might interpret the key idea of the universe. In this conversation, we will thoroughly analyze three unmistakable translations of quantum mechanics: the Copenhagen Understanding, the Many-Universes Speculation, and a determination of different understandings that give elective points of view on this mysterious field.

The Copenhagen Translation, figured out by Niels Bohr and Werner Heisenberg during the 1920s, is one of the most broadly perceived and shown understandings of quantum mechanics. As indicated by this view, the quantum world is innately probabilistic. A quantum framework is portrayed by a numerical element known as the wave capability, and it exists in a superposition of potential states until an estimation is made.

Right now of estimation, the wave capability falls to yield a particular result, and the demonstration of estimation assumes a focal part in the hypothesis. Until noticed, quantum frameworks don't have distinct properties.

The Copenhagen Understanding has been exceptionally fruitful in making sense of exploratory outcomes and giving a down to earth structure to quantum physical science. It puts critical accentuation on the job of the eyewitness and the inborn indeterminacy of quantum frameworks, an idea typified by Heisenberg's Vulnerability Guideline. This guideline directs that specific sets of properties, like an electron's situation and energy, can't be at the same time estimated with erratic accuracy.

Nonetheless, the Copenhagen Translation has additionally created significant philosophical inquiries and reactions. One of the essential worries is that it neglects to give an unmistakable ontological structure to figuring out the idea of the quantum world. It brings up issues about what befalls a quantum framework before it is estimated and whether there is a hidden reality that exists freely of perception. This understanding, with its emphasis on the spectator and the breakdown of the wave capability, stays a subject of dynamic discussion inside the logical and philosophical networks.

As opposed to the Copenhagen Understanding, the Many-Universes Speculation, proposed by Hugh Everett III in 1957, offers an emphatically alternate point of view. It recommends that each conceivable result of a quantum estimation really happens, however every result exists in a different, non-conveying part of the universe. In this view, the universe continually parts into various equal real factors, and all possible results coincide in various branches. The Many-Universes Speculation gives a deterministic system where the wave capability never implodes, and the onlooker follows one of various potential ways inside a consistently extending multiverse.

The Many-Universes Speculation has acquired consideration among a physicists and offers an answer for the "estimation issue" inborn in the Copenhagen Understanding.

Nonetheless, it presents its own arrangement of philosophical requests, for example, the idea of these equal real factors and the mechanics of universe parting. The possibility that each conceivable quantum result has a comparing branch in an equal universe has significant ramifications for how we might interpret the idea of the real world and the job of the onlooker.

Other than the Copenhagen Understanding and the Many-Universes Speculation, various different translations of quantum mechanics have been proposed. We should momentarily investigate a portion of these elective perspectives:

Pilot-Wave Hypothesis (Bohmian Mechanics): Created by Louis de Broglie and David Bohm during the 1950s, this translation proposes that quantum particles are directed by a hid pilot wave that decides their directions.

As opposed to the Copenhagen Understanding, which depicts specific properties as essentially questionable until estimated, Bohmian mechanics presents a deterministic structure where the molecule's situation and energy are represented by the wave capability and beginning circumstances. While it gives an unmistakable ontological system, it brings up issues about the idea of the secret factors and presents the idea of non-region, where particles can momentarily impact each other's way of behaving a ways off, testing the standards of extraordinary relativity.

Conditional Understanding: Proposed by John Cramer during the 1980s, this translation sees quantum communications as a two-step process including both high level (future-coordinated) and hindered (past-coordinated) waves. The Conditional Understanding proposes that quantum exchanges include correspondence through these waves, making a standing wave that addresses an exchange between the producer and safeguard. While it offers an interesting method for understanding quantum trap and the breakdown of the wave capability, it isn't quite so broadly embraced as different translations and stays a subject of progressing examination and discussion.

Objective Breakdown Models: These models suggest that the wave capability unexpectedly falls because of an actual interaction, free of estimation. One noticeable model is the Persistent Unconstrained Confinement (CSL) hypothesis, which places that particles experience irregular, ceaseless breakdowns, prompting a dissemination of their wave capabilities. Objective Breakdown Models mean to overcome any barrier between the quantum and traditional universes and proposition testable expectations through tests intended to distinguish unconstrained breakdowns.

Social Translation: Presented via Carlo Rovelli, the Social Understanding difficulties the thought of an objective reality and underlines the social idea of actual peculiarities. As per this view, properties of a quantum object are characterized corresponding to another item or eyewitness, and there are no outright properties or states. This understanding lines up with a more extensive philosophical pattern known as relationalism, which questions the idea of outright existence.

Quantum Bayesianism (QBism): QBism is a translation that puts the onlooker at the focal point of quantum hypothesis. It recommends that the quantum wave capability addresses a specialist's emotional convictions about a quantum framework's results. In this view, quantum mechanics is viewed as a system for refreshing one's very own levels of conviction in light of proof and perceptions. QBism offers an extreme change in context, underscoring that quantum mechanics is a hypothesis about the individual encounters and convictions of eyewitnesses.

Many-Personalities Understanding: Proposed by Adrian Kent, the Many-Personalities Translation shares an emphasis on the eyewitness' job. In this view, every estimation result makes a new "mind," and the quantum wave capability addresses the dissemination of psyches in various states. The Many-Personalities Understanding proposes that every eyewitness sees a solitary, distinct result, while all potential results exist in the "scene of psyches."

These elective translations challenge the basic suppositions and ideas of quantum mechanics in different ways. They bring up issues about the idea of the real world, the job of estimation, and the connection between the quantum and old style universes. While certain understandings stand out and uphold than others, recognizing the variety of viewpoints inside the field of quantum mechanics is fundamental. Every understanding offers a one of a kind focal point through which to see the secretive and complex quantum domain.

The continuous discussions and conversations encompassing these translations highlight the significant idea of quantum mechanics and its capacity to ignite interest, investigation, and development. As how we might interpret the quantum world develops and as new exploratory proof opens up, these translations will keep on advancing, offering new bits of knowledge into the central activities of the universe. Whether one lines up with the Copenhagen Translation, the Many-Universes Speculation, or any of different understandings, the investigation of quantum mechanics stays a spellbinding excursion into the core of the normal world's most confounding domain.

8.3 Ongoing debates and questions about the nature of quantum reality

The universe of quantum mechanics is a domain of profound secrets and baffling peculiarities that keep on testing how we might interpret the essential idea of the universe. While quantum mechanics gives a strong numerical system to portraying the way of behaving of particles at the infinitesimal level, the understandings and questions encompassing it stay a subject of extreme discussion and investigation. In this conversation, we will dive into the continuous discussions and questions that spin around the idea of quantum the truth, its philosophical ramifications, and the quest for a thorough comprehension of the quantum world.

The Estimation Issue: The estimation issue is a focal puzzle in quantum mechanics. It emerges from the clear wave-molecule duality of quantum frameworks. When unnoticed, particles exist in a superposition of potential states, depicted by a wave capability. In any case, upon estimation, the wave capability falls to a particular result. The inquiry is, what causes this breakdown, and how does estimation influence the way of behaving of quantum frameworks? The Copenhagen Translation states that estimation assumes a basic part, while different understandings, similar to the Many-Universes Speculation and Goal Breakdown Models, propose elective answers for this issue.

The Idea of Wave Works: The wave capability is a major idea in quantum mechanics, portraying the quantum condition of a framework. Be that as it may, its careful nature and understanding stay a subject of discussion. Is the wave capability only a numerical device for making expectations, as certain understandings recommend, or does it address an actual element that depicts the genuine condition of a quantum framework? This question prompts the discussion about whether the wave capability is an ontological or epistemological idea.

Quantum Non-region: Quantum snare, a peculiarity where the properties of particles become related, in any event, when isolated by tremendous distances, has confounded physicists since its disclosure. This non-neighborhood relationship appears to resist old style instincts about causality and area. While the infringement of Ringer's imbalances tentatively backings the truth of trap, the philosophical inquiries encompassing the idea of non-area continue. How do entrapped particles "impart" quicker than the speed of light, or is there a fundamental system that makes sense of this peculiarity?

Determinism versus Indeterminism: Quantum mechanics brings a degree of indeterminacy into the major texture of the universe. In numerous understandings, including the Copenhagen Translation, the way of behaving of quantum frameworks shows up innately probabilistic, opposing old style determinism. In any case, translations like Bohmian mechanics offer deterministic other options. The discussion among determinism and indeterminism brings up issues about whether the quantum world is in a general sense unusual or on the other hand assuming there are covered up factors that decide results.

The Job of the Onlooker: The Copenhagen Translation underscores the job of the spectator in quantum mechanics, recommending that estimation is a characterizing factor in the way of behaving of quantum frameworks. This perspective has significant ramifications for the idea of the real world and the job of awareness in molding the quantum world. Different translations, like QBism and the Many-Personalities Understanding, place the spectator at the focal point of quantum hypothesis in various ways, testing conventional ideas of objectivity and outside the real world.

Quantum Authenticity versus Against Authenticity: Quantum authenticity alludes to the conviction that quantum frameworks have objective properties and that the quantum world has a hidden, free reality. Conversely, hostile to authenticity declares that quantum mechanics is a structure for depicting emotional encounters and doesn't be guaranteed to mirror an outer reality. The discussion among authenticity and against authenticity addresses the essential inquiry of how it affects something to be genuinely with regards to the quantum domain.

Breakdown of the Wave Capability: The idea of wave capability breakdown, fundamental to numerous understandings, brings up issues about what happens when a quantum framework is estimated. Does the wave work breakdown to a solitary positive state, as proposed by the Copenhagen Understanding, or do all potential results exist at the same time in various parts of the Many-Universes Speculation? The idea of the breakdown and the ramifications for the presence of equal real factors remain subjects of progressing conversation.

Development of Traditional from Quantum: A captivating inquiry in the way of thinking of quantum mechanics is the manner by which the old style world, portrayed by distinct and obvious properties, rises up out of the quantum domain, where frameworks can exist in superpositions.

This cycle, frequently alluded to as quantum-to-old style change, isn't completely perceived. Analysts investigate different models and speculations to explain this progress and the limit between the quantum and traditional universes.

Quantum Data and Calculation: Quantum mechanics has brought about the field of quantum data hypothesis, which investigates the utilization of quantum frameworks for data handling and calculation. Quantum PCs, outfitting the one of a kind properties of quantum states, can possibly change processing. Notwithstanding, questions persevere with respect to a definitive cutoff points and potential outcomes of quantum data handling, as well as its ramifications for how we might interpret the idea of calculation and reality.

Trial of Central Physical science: Exploratory trial of crucial physical science, for example, Ringer tests and examinations looking at the infringement of Chime's disparities, assume a urgent part in our investigation of the quantum world. These investigations try to affirm or challenge the forecasts of quantum mechanics and its translations. Continuous examination keeps on pushing the limits of our insight and may prompt new experiences into the idea of quantum reality.

Quantum Advances: Quantum mechanics has pragmatic applications, prompting the improvement of quantum innovations like quantum cryptography and quantum sensors. The proceeded with headway of these advances offers potential chances to test the quantum domain and test its limits, bringing up issues about what these innovations might mean for how we might interpret quantum mechanics and the idea of the real world.

Exchange with Other Logical Speculations: The connection between quantum mechanics and other logical hypotheses, like general relativity with regards to quantum gravity, stays a significant open inquiry. Coordinating the standards of quantum mechanics with those of general relativity presents huge difficulties and could prompt new bits of knowledge into the idea of the universe.

Translation Variety: The huge number of understandings of quantum mechanics, each offering an alternate point of view on quantum reality, features the variety of thoughts and perspectives inside the logical and philosophical networks. The continuous discussion about which understanding best catches the real essence of quantum reality mirrors the intricacy and profundity of the subject.

All in all, the idea of quantum reality stays a rich and puzzling area of investigation. The continuous discussions and questions encompassing quantum mechanics address the most principal parts of the universe — inquiries regarding the idea of the real world, the job of estimation, and the connection between the quantum and traditional universes.

These requests keep on moving interest, challenge our predispositions, and drive forward how we might interpret the quantum domain. As scientists and rationalists dive further into the secrets of quantum mechanics, new experiences and disclosures anticipate, promising to reveal insight into the puzzling idea of the quantum universe and shape the fate of science and reasoning.

Chapter 9

Unresolved Mysteries and the Future of Quantum Reality

The universe of quantum material science has, for more than 100 years, interested and frustrated researchers and masterminds the same. It is a domain where particles can exist in different states all the while, and where apparently perplexing peculiarities challenge our essential comprehension of the universe. Quantum mechanics, as a hypothesis, has substantiated itself strikingly fruitful in portraying the way of behaving of the littlest constituents of our universe, but it leaves us with a plenty of unsettled secrets and inquiries concerning the idea of reality itself. In this exposition, we will investigate a portion of these unsettled secrets and examine the eventual fate of quantum reality, a future that guarantees both more profound experiences and new, unfamiliar regions of understanding.

One of the most confusing parts of quantum mechanics is the idea of superposition. Superposition permits particles to exist in various states without a moment's delay. A molecule can be both a wave and a molecule all the while until it is estimated. This confounding idea challenges our old style instinct, which depends on the possibility that items ought to have unequivocal properties consistently. The well known twofold cut analyze shows this peculiarity clearly. At the point when particles are sent through two cuts, they act as the two waves and particles, making an obstruction design on a screen. In any case, when we see what cut the molecule goes through, it out of nowhere acts as a solitary molecule. This progress from superposition to an unequivocal state upon perception is a secret that stays unsettled.

One more puzzle of quantum mechanics is snare. Ensnarement happens when two particles become corresponded so that the condition of one molecule quickly impacts the condition of the other, no matter what the distance isolating them. Albert Einstein broadly alluded to this as "creepy activity a ways off." The idea of snare difficulties how we might interpret region and the speed of data move, as it strongly implies that data can travel quicker than the speed of light. While trap has been tentatively affirmed, we are yet to get a handle on the hidden systems and ramifications of this peculiarity completely.

Quantum vulnerability is one more unsettled secret at the core of quantum mechanics. Heisenberg's vulnerability standard expresses that specific sets of properties, like position and force, can't be exactly known at the same time. The more precisely we measure one property, the less precisely we can know the other. This standard brings an inborn eccentricism into the quantum world, testing the determinism of traditional physical science. While the vulnerability standard is a key part of quantum mechanics, its philosophical and mystical ramifications stay a subject of discussion.

The estimation issue is one more vexing issue in quantum mechanics. It connects with the demonstration of estimation itself and the breakdown of the wave capability. At the point when a quantum framework is estimated, its superposition implodes into a solitary, unmistakable state. In any case, the idea of this breakdown is as yet a question of discussion. A few translations of quantum mechanics, like the Copenhagen understanding, propose that the demonstration of estimation assumes a major part in deciding the result. Others, similar to the many-universes translation, suggest that all potential results happen in isolated parts of the universe, making a multiverse. Settling the estimation issue is urgent for a far reaching comprehension of the quantum world.

Quantum gravity is one more boondocks in the investigation of the quantum domain. General relativity, which depicts gravity for a vast scope, and quantum mechanics, which oversees the way of behaving of particles on the littlest scales, are two of the best hypotheses in material science. Be that as it may, they are generally contradictory. Consolidating these two speculations into a solitary, intelligent structure — a hypothesis of quantum gravity — stays a significant test. Such a hypothesis would be fundamental for figuring out the way of behaving of issue and spacetime at the smallest scales, like nearby dark openings or during the early snapshots of the universe.

Dim matter and dim energy, two secretive parts that make up most of the universe, additionally meet with quantum material science. While dim matter's presence is deduced from its gravitational consequences for apparent matter, its real essence stays obscure. Quantum mechanics might actually give experiences into the idea of dim matter, as it could include new sorts of particles past the Standard Model of molecule physical science.

Essentially, dull energy, liable for the sped up extension of the universe, represents a central test to how we might interpret gravity and spacetime. A quantum hypothesis of gravity might reveal insight into the idea of dull energy.

Quantum registering is a region where the secrets of quantum mechanics hold both commitment and interest. Quantum PCs influence the standards of superposition and trap to perform calculations that would be basically outside the realm of possibilities for old style PCs. These machines can possibly change fields like cryptography, materials science, and medication disclosure. Notwithstanding, building down to earth quantum PCs and bridling their power is an imposing specialized challenge that requires defeating issues connected with quantum decoherence and mistake revision.

The eventual fate of quantum the truth is one of progressing investigation and disclosure. While these unsettled secrets present critical difficulties, they additionally offer open doors for leap forwards that could reform how we might interpret the universe and lead to mechanical progressions with significant ramifications. Scientists are effectively attempting to foster new trials and speculations that might give replies to these mysteries, like the advancement of additional exact estimations, novel ways to deal with quantum gravity, and the investigation of elective translations of quantum mechanics.

All in all, the universe of quantum physical science is a domain of significant secrets that keep on enamoring researchers and logicians the same. Superposition, trap, vulnerability, and the estimation issue challenge our old style instincts and push the limits of our comprehension. Quantum gravity, dull matter, and dim energy suggest key conversation starters about the idea of the universe, while quantum registering opens up additional opportunities for calculation and innovation. The eventual fate of quantum reality holds the commitment of more profound experiences into the idea of the universe, yet it additionally helps us to remember the significant secrets that keep on molding our mission for information. The quest for replies to these inquiries will without a doubt prompt new ideal models and groundbreaking disclosures in the next few decades, molding the fate of science and our comprehension of the universe.

9.1 Remaining enigmas in the quantum realm

The quantum domain, that strange space overseeing the way of behaving of particles at the littlest scales, proceeds to astound and challenge how we might interpret the major idea of the real world. Starting from the commencement of quantum mechanics in the mid twentieth 100 years, researchers have taken striking steps in making sense of and foreseeing the way of behaving of subatomic particles. In any case, underneath the obvious clearness and progress of quantum hypothesis, there stay various conundrums and unsettled questions that have fascinated and puzzled physicists and logicians the same. In this paper, we will dig into a portion of these leftover conundrums in the quantum domain, investigating the secrets that keep on inciting interest and drive logical request.

Quite possibly of the most basic riddle in quantum mechanics is the idea of wave-molecule duality. This baffling thought places that particles, for example, electrons and photons, display both wave-like and molecule like ways of behaving, contingent upon how they are noticed or estimated. For example, in the well known twofold cut try, when electrons are terminated at an obstruction with two cuts, they act as though they are waves, making an impedance design on the locator screen. In any case, when an estimation is made to figure out what cut every electron goes through, they act as particles and produce a particular example on the screen. This peculiarity brings up the issue of whether particles have inherent properties or whether their inclination is setting reliant, a problem that still can't seem to be completely settled.

The idea of quantum entrapment remains as one more problem that has captivated and beguiled researchers for quite a long time. Entrapment happens when at least two particles become corresponded so that the condition of one molecule is immediately connected to the condition of another, no matter what the spatial partition between them. This non-nearby association appears to challenge old style thoughts of causality and region, driving Albert Einstein to broadly allude to it as "creepy activity a good ways off." While exploratory proof emphatically upholds the truth of ensnarement, the basic systems and the speed at which data is traded between trapped particles stay open inquiries.

Quantum indeterminacy, as exemplified in Heisenberg's vulnerability guideline, is one more charming riddle in quantum mechanics. This rule places that specific sets of corresponding properties, like a molecule's situation and force, can't be unequivocally known all the while. The more precisely we measure one property, the less unequivocally we can know the other. This innate vulnerability challenges the old style thought of a deterministic universe where exact expectations are conceivable. The philosophical ramifications of indeterminacy keep on igniting discusses in regards to the idea of the real world and the job of estimation in quantum frameworks.

The estimation issue is a lasting wellspring of disarray and discussion in quantum mechanics. It spins around the demonstration of estimation itself and the breakdown of the wave capability, which depicts the likelihood dispersion of a quantum framework's potential states. At the point when an estimation is made, the wave capability implodes into a particular state, and the framework is said to have a clear incentive for the deliberate property. In any case, the idea of this breakdown stays quite possibly of the most petulant issue in quantum physical science. A few understandings of quantum mechanics, for example, the Copenhagen translation, set that estimation assumes an essential part in deciding the result. Others, similar to the many-universes translation, suggest that all potential results happen in discrete parts of the universe, making a multiverse of real factors. Settling the estimation issue is fundamental for a complete comprehension of quantum physical science.

The journey for a hypothesis of quantum gravity is one more unsettled puzzler in the quantum domain. General relativity, Albert Einstein's hypothesis of gravity, depicts the power that oversees the way of behaving of issue and spacetime on astronomical scales. Quantum mechanics, then again, gives a strikingly effective structure to figuring out the way of behaving of particles on microscopic scales. However, these two mainstays of present day physical science are in a general sense contradictory. Joining them into a solitary, cognizant hypothesis of quantum gravity is an imposing test, one that is fundamental for grasping the way of behaving of issue and spacetime under outrageous circumstances, like those tracked down close to dark openings or during the early snapshots of the universe.

The baffling peculiarities of dull matter and dim energy additionally cross with the quantum domain. Dull matter is a baffling and imperceptible type of issue that applies gravitational effect on noticeable matter, yet it doesn't communicate through electromagnetic powers. Its real essence stays obscure, yet including intriguing particles past the Standard Model of molecule physics is accepted. Quantum mechanics might hold the way to grasping the key properties of dull matter. Dull energy, which is liable for the sped up extension of the universe, acquaints a significant test with how we might interpret gravity and spacetime. A quantum hypothesis of gravity might actually reveal insight into the idea of dim energy and its suggestions for the universe.

The arising field of quantum registering addresses a tempting riddle with the possibility to change different logical and innovative spaces. Quantum PCs influence the standards of superposition and ensnarement to perform calculations that would be restrictively tedious or inconceivable for old style PCs. These machines can possibly decisively speed up progress in fields like cryptography, materials science, and medication disclosure. Be that as it may, building pragmatic and versatile quantum PCs stays a considerable specialized challenge. Beating issues connected with quantum decoherence and mistake amendment is urgent for saddling the full force of quantum calculation.

The eventual fate of quantum the truth is one of continuous investigation and disclosure. While these unsettled puzzlers present critical difficulties, they likewise offer open doors for forward leaps that could reform how we might interpret the universe and lead to mechanical headways with significant ramifications. Analysts are effectively chipping away at the improvement of new trials and hypotheses that might give replies to these riddles. For instance, the headway of quantum innovations, like more exact estimations and the investigation of elective understandings of quantum mechanics, holds the commitment of revealing insight into the idea of wave-molecule duality and the job of estimation.

Quantum data science and quantum figuring research are blossoming fields that may ultimately disentangle the secrets of quantum trap and quantum indeterminacy. Tests including ensnared particles, for example, those directed in quantum instant transportation and quantum key dissemination, keep on pushing the limits of how we might interpret non-region and the security of correspondence.

Quantum figuring, while still in its outset, has gained critical headway, and the improvement of quantum blunder rectifying codes holds the possibility to make huge scope quantum PCs down to earth.

Investigating the idea of quantum gravity stays a focal mission in contemporary physical science. String hypothesis, circle quantum gravity, and other quantum gravity approaches are effectively sought after, offering various systems for accommodating the quantum and gravitational domains. These hypotheses plan to give a bound together depiction of the powers overseeing the universe, from the subatomic to the enormous scope. Progress in this attempt wouldn't just purpose the quantum gravity puzzler yet in addition shed light on the idea of dull matter and dim energy.

All in all, the quantum domain stays a space of significant puzzles and secrets that keep on enamoring researchers, scholars, and the inquisitive the same. The puzzle of wave-molecule duality challenges our old style instincts, while quantum entrapment brings up issues about the central idea of the real world and the restrictions of how we might interpret causality. Quantum indeterminacy challenges the deterministic perspective that has molded traditional material science for a really long time, and the estimation issue highlights the confounding job of the onlooker in quantum frameworks.

The quest for a hypothesis of quantum gravity, the idea of dull matter and dim energy, and the improvement of quantum registering innovations address progressing missions to open the mysteries of the quantum domain. While these puzzlers present critical difficulties, they likewise offer tempting open doors for more profound bits of knowledge, outlook changing revelations, and extraordinary headways in science and innovation. The eventual fate of quantum the truth is one of proceeded with investigation and an update that the universe, at its generally central level, stays a wellspring of marvel, interest, and ceaseless secret.

9.2 The prospects for future discoveries and breakthroughs

The journey for information and figuring out has been a principal trait of humankind over the entire course of time. From the earliest logical requests to the latest mechanical headways, the quest for disclosure and forward leaps has driven progress and development. In this exposition, we will investigate the possibilities for future disclosures and leap forwards across an extensive variety of logical, mechanical, and cultural spaces, taking into account both the known outskirts of investigation and the strange regions that entice us.

Space Investigation:

One of the most remarkable outskirts for future disclosures is space investigation. Human interest in the universe has driven us to investigate our nearby planet group and then some. Mechanical missions to Mars have uncovered interesting traces of past water streams and raised the enticing chance of past or present microbial life. What's to come holds the commitment of sending people to Mars, perhaps inside the following years and years, offering the chance for exceptional disclosures and the expected affirmation of extraterrestrial life.

Past our planetary group, the quest for exoplanets — planets outside our nearby planet group — is a blossoming field. The revelation of Earth-like exoplanets in the tenable zone of different stars would be a stupendous forward leap, possibly meaning the presence of life past Earth. Strong telescopes, for example, the James Webb Space Telescope, are not too far off and will empower us to peer further into the universe and study exoplanets with more prominent accuracy.

Molecule Material science:

Molecule material science keeps on pushing the limits of how we might interpret the principal building blocks of the universe. The Huge Hadron Collider (LHC) at

CERN has proactively affirmed the presence of the Higgs boson, a molecule that gives mass to different particles. Be that as it may, many inquiries stay unanswered, like the idea of dim matter, dim energy, and the unification of the principal powers. Future molecule gas pedals, similar to the proposed High-Radiance LHC, could give bits of knowledge into these secrets and uncover altogether new particles and peculiarities.

Quantum Registering and Data:

The field of quantum registering holds the commitment of groundbreaking leap forwards. Quantum PCs tackle the standards of superposition and entrapment to perform computations at speeds that old style PCs can't coordinate. While reasonable quantum PCs are still in the beginning phases of improvement, their potential applications are faltering. They could reform cryptography, drug revelation, advancement issues, and materials science. The following couple of many years might see huge improvement in building adaptable and mistake remedied quantum PCs.

Quantum data science, which envelops quantum cryptography and quantum correspondence, additionally can possibly reshape the field of secure data trade. The advancement of quantum-safe encryption and secure quantum correspondence organizations will be fundamental to safeguard our computerized framework later on.

Biotechnology and Medication:

In biotechnology and medication, the possibilities for leap forwards are enormous. Hereditary designing, especially through CRISPR-Cas9 innovation, offers the capacity to alter the DNA of life forms, including people. While this can possibly treat hereditary illnesses, it additionally brings up moral and cultural issues that request cautious thought.

Regenerative medication, including foundational microorganism treatment and tissue designing, holds the commitment of fixing or supplanting harmed or deteriorated tissues and organs. Headways in customized medication and designated treatments might prompt more viable medicines with less aftereffects.

The mission for a more profound comprehension of the human cerebrum proceeds, with drives like the Human Mind Task and the Mind Drive in the US. Planning the brain circuits and elements of the mind could prompt forward leaps in neuroscience, man-made reasoning, and our comprehension of cognizance.

Environmental Change and Manageability:

Tending to environmental change and manageability is one of the most squeezing difficulties within recent memory. Examination into sustainable power sources, like high level sunlight based chargers, energy capacity, and cutting edge atomic reactors, offers the potential for leap forwards in diminishing fossil fuel byproducts and alleviating the impacts of a worldwide temperature alteration.

The improvement of carbon catch and usage innovations could give a fundamental device in eliminating overabundance carbon dioxide from the air. Furthermore, progressions in manageable agribusiness and hydroponics, as well as developments in water decontamination and protection, will be vital in satisfying the world's developing needs for food and freshwater.

Computerized reasoning and AI:

Computerized reasoning (simulated intelligence) and AI are quickly developing fields with critical possibilities for future revelations and leap forwards. Profound learning, a subfield of AI, has exhibited striking triumphs in picture and discourse acknowledgment, normal language handling, and playing complex games like Go and chess. These advances have applications in fields as different as medical services, finance, and independent vehicles.

Quantum AI is a promising area of exploration, where quantum PCs can be utilized to speed up AI calculations, possibly taking care of intricate issues quicker and all the more proficiently. The convergence of quantum figuring and computer based intelligence could prompt earth shattering improvements in drug disclosure, materials science, and enhancement issues.

Sustainable power and Natural Innovation:

The change to a manageable and low-carbon energy future is a worldwide objective. The improvement of cutting edge sustainable power innovations, like more proficient and financially savvy sunlight based chargers, high level breeze turbines, and cutting edge batteries, can possibly alter the manner in which we create and store energy.

Developments in natural innovation, including air and water cleansing frameworks, squander the executives, and feasible structure materials, can assist with relieving the ecological difficulties we face. Forward leaps here can essentially decrease contamination and work on the personal satisfaction for individuals all over the planet.

Nanotechnology:

Nanotechnology, the control of issue at the nanoscale, offers energizing possibilities for leap forwards in materials science, hardware, and medication. By designing materials at the nuclear and sub-atomic levels, we can make new materials with unprecedented properties. These materials can be utilized in everything from super-solid and lightweight composites to profoundly effective sun oriented cells.

In medication, nanotechnology holds the possibility to upset drug conveyance, empowering designated treatments with decreased secondary effects. Nanoscale sensors and gadgets can give uncommon experiences into natural cycles and consider early sickness identification.

Astronomy and Cosmology:

How we might interpret the universe's starting point and development keeps on being a wellspring of interest and revelation. Perceptions of infinite microwave foundation radiation and the dissemination of systems have given vital experiences into the early universe. Future missions, for example, the James Webb Space Telescope and the Square Kilometer Exhibit, are ready to uncover more about the starting points of worlds, the idea of dull matter and dim energy, and the expected presence of extra-terrestrial life.

Neuroscience and Mind PC Connection points:

The investigation of the human mind and the advancement of cerebrum PC interfaces (BCIs) are fields with significant ramifications for future revelations. BCIs can possibly change the existences of people with incapacities, permitting them to impart, control prosthetic gadgets, and recover a degree of freedom beforehand impossible.

Further progressions in understanding the cerebrum's complexities and growing more modern BCIs could open new wildernesses in human-PC association and upgrade mental capacities. In any case, the moral and cultural ramifications of mind PC connection points will require cautious thought and guideline.

Materials Science and High level Assembling:

Materials science assumes a vital part in numerous mechanical headways. The revelation of novel materials with striking properties can prompt leap forwards in gadgets, transportation, and energy stockpiling. The advancement of 2D materials, for example, graphene, and the investigation of topological materials have previously shown extraordinary commitment.

High level assembling strategies, including 3D printing and added substance fabricating, offer the possibility to alter creation processes, lessening waste and empowering more modified and productive assembling. The continuous investigation of nanomaterials and metamaterials could prompt materials with remarkable properties, for example, imperceptibility shrouds or materials that can control light and sound in novel ways.

Natural Protection and Biodiversity:

Preservation endeavors and biodiversity research are basic for protecting the planet's environments and forestalling species terminations. Progresses in preservation science, including the utilization of DNA barcoding and genomics, can help distinguish and safeguard imperiled species all the more really. Besides, developments in natural surroundings reclamation and preservation innovation can support protecting delicate environments.

Cultural and Moral Contemplations:

As we adventure into these strange domains of disclosure, taking into account the cultural and moral ramifications of our breakthroughs is fundamental. Moral predicaments will emerge in fields like biotechnology, artificial intelligence, and hereditary qualities, where the ability to control life and awareness has significant outcomes. Resolving inquiries of value, access, and the mindful utilization of arising advancements will be fundamental to guaranteeing that our revelations .

9.3 The enduring impact of quantum reality on science and society

The appearance of quantum mechanics in the mid twentieth century changed how we might interpret the major structure blocks of the universe. Quantum hypothesis, with its mysterious standards and confusing peculiarities, enduringly affects both established researchers and society at large. In this exposition, we will investigate the persevering through impact of quantum reality, looking at its ramifications for science, innovation, reasoning, and society.

Logical Change in perspective:

Quantum mechanics started a change in perspective in science. It tested old style material science, which had ruled for quite a long time, and presented another system for figuring out the way of behaving of particles on the littlest scales. The standards of quantum hypothesis, like superposition, ensnarement, and wave-molecule duality, challenged old style instinct and required an extremist reexamining of the idea of the real world.

Quantum mechanics gave a strikingly effective portrayal of the way of behaving of subatomic particles. It made sense of peculiarities, for example, the quantization of energy levels in particles, the discrete ghostly lines in nuclear spectra, and the dependability of issue. This achievement supported trust in the logical strategy and showed the way that how we might interpret the universe could advance and adjust to new revelations.

Mechanical Progressions:

The effect of quantum mechanics on innovation has been groundbreaking. Quantum hypothesis supports the advancement of numerous cutting edge innovations, from semiconductors and lasers to X-ray machines and the worldwide situating framework (GPS).

Semiconductor gadgets, including semiconductors and diodes, depend on the standards of quantum mechanics to work. These gadgets structure the foundation of present day hardware, empowering the formation of PCs, cell phones, and incalculable different devices that have reformed correspondence and data handling.

Lasers, which are necessary to many applications, work on quantum standards. They have empowered forward leaps in fields like broadcast communications, clinical medical procedure, and accuracy estimations. Quantum specks, a quantum mechanical peculiarity, have prompted developments in show innovation, working on the nature of TV screens and PC screens.

Attractive reverberation imaging (X-ray) machines are fundamental for painless clinical diagnostics. They depend on the quantum properties of nuclear cores to make point by point pictures of interior body structures. In the field of route, the GPS framework relies upon the exact synchronization of nuclear clocks, which, thusly, depends on the standards of quantum mechanics.

Quantum Figuring:

Quantum figuring addresses perhaps of the most encouraging boondocks in innovation. Quantum PCs influence the standards of superposition and ensnarement to perform computations at speeds that traditional PCs can't achieve. While viable quantum PCs are still in the beginning phases of improvement, they hold the possibility to alter different fields.

Quantum PCs can tackle complex issues, like calculating enormous numbers, recreating quantum frameworks, and streamlining arrangements, a lot quicker than old style PCs. These capacities have significant ramifications for cryptography, drug disclosure, and materials science. For example, quantum PCs might actually break current encryption techniques, requiring the improvement of quantum-safe cryptography.

The improvement of versatile and blunder revised quantum PCs is a significant specialized challenge, yet it holds the commitment of resolving issues that were beforehand obstinate. Quantum calculations, like Shor's calculation for figuring enormous numbers and Grover's calculation for looking through unsorted data sets, have shown the capability of quantum processing to upset existing innovations.

Quantum Correspondence:

Quantum mechanics has additionally brought about quantum correspondence, which is possibly the most reliable type of correspondence in presence. Quantum key dispersion (QKD) permits two gatherings to share cryptographic keys in a manner that is safe to listening in. Any endeavor to capture the quantum key would unavoidably modify its quantum state, making the imparting parties aware of potential security breaks.

Quantum correspondence advancements have applications in secure military correspondences, monetary exchanges, and safeguarding touchy data. Quantum encryption might turn out to be progressively significant in a time where information security is of fundamental concern.

Philosophical Ramifications:

The philosophical ramifications of quantum mechanics have ignited persevering through discussions and conversations. The indeterminacy presented by Heisenberg's vulnerability rule difficulties the traditional idea of a deterministic universe. It infers that at the quantum level, a few properties of particles can't be definitively known, and the result of estimations is innately dubious.

The estimation issue in quantum mechanics, which manages the idea of wave capability breakdown and the job of perception, has been the subject of continuous philosophical request. Different translations of quantum mechanics, including the Copenhagen understanding, many-universes translation, and goal breakdown models, offer alternate points of view on this principal issue.

The way of thinking of quantum mechanics reaches out to inquiries concerning the idea of the real world, determinism, and the job of the eyewitness in forming quantum frameworks. These philosophical conversations keep on affecting the manner in which we contemplate the idea of the universe and our place inside it.

Cultural Effects:

The cultural effects of quantum mechanics are complex. Mechanical headways driven by quantum hypothesis have changed daily existence. The omnipresence of electronic gadgets, lasers, and correspondence advances has reshaped the manner in which individuals live and work. Quantum innovations play had a urgent impact in current medical services, empowering clinical imaging, exact diagnostics, and therapies.

In the domain of safety and guard, quantum advances have significant ramifications. Quantum correspondence and encryption can possibly defend touchy data and interchanges, which is critical during a time of computerized data sharing and digital dangers. Quantum advancements are ready to assume a fundamental part in safeguarding public safety and protection.

The getting through effect of quantum mechanics additionally reaches out to schooling and exploration. The investigation of quantum mechanics has turned into a foundation of current physical science schooling, and it is fundamental for understudies and specialists in fields as different as materials science, science, and software engineering. As quantum advances keep on propelling, the interest for specialists in quantum mechanics and quantum designing is probably going to develop.

Challenges and Moral Contemplations:

In spite of its various advantages, the getting through effect of quantum mechanics likewise presents difficulties and moral contemplations. The advancement of quantum innovations, especially quantum registering, raises worries about network safety. Quantum PCs might actually break existing encryption strategies, prompting security weaknesses.

The moral utilization of quantum innovations, like genome altering methods in biotechnology, is a subject of continuous discussion. Advances like CRISPR-Cas9 empower the control of DNA, bringing up issues about the possible results, moral limits, and guideline of hereditary adjustments in people and different life forms.

Quantum correspondence advancements, while exceptionally secure, could likewise be utilized for pernicious purposes. Guaranteeing that quantum advances are utilized dependably and morally is urgent to forestall abuse and unseen side-effects.

Quantum Reality and the Unknown Future:

Quantum reality stays a wellspring of marvel and investigation, with numerous puzzles and unsettled questions. The idea of wave-molecule duality, quantum entrapment, the estimation issue, and quantum gravity keep on dazzling researchers and masterminds. The journey for a brought together hypothesis that accommodates quantum mechanics with general relativity, known as a hypothesis of quantum gravity, stays a focal test in contemporary physical science.

As we plan ahead, quantum mechanics will probably keep on motivating leap forwards in science and innovation. Quantum registering, quantum correspondence, and quantum data science address fields with huge possibilities for disclosures that could upset how we might interpret the world and the instruments we use to explore it.

All in all, the persevering through effect of quantum reality on science and society is a demonstration of the getting through force of human interest and logical request. From innovative headways and philosophical conversations to moral contemplations and progressing difficulties, quantum mechanics has made a permanent imprint on our reality. As we adventure into the unknown domain of the quantum domain, we have the open door and obligation to bridle its true capacity to improve mankind while resolving the complex moral and cultural inquiries it raises. Quantum reality keeps on directing our investigation of the universe's most significant secrets, guaranteeing that the tradition of quantum mechanics perseveres into what's in store.